the perfect
puppy

the perfect puppy

GWEN BAILEY

hamlyn

For Spider, Winnie, Beau, Sammy and Scampi – the lovely dogs I have shared my life with, who have given so much love and happiness and taught me more than I could have imagined.

An Hachette UK Company
www.hachette.co.uk

This revised edition first published in Great Britain in 2008 by Hamlyn, a division of
Octopus Publishing Group Ltd
Carmelite House, 50 Victoria Embankment
London, EC4Y 0DZ
www.octopusbooks.co.uk

Originally published in Great Britain in 1995

ISBN 978-0-600-61722-8

A CIP catalogue record for this book is available from the British Library.

Printed and bound in China

The advice in this book is provided as general information only. It is not necessarily specific to any individual case and is not a substitute for the guidance and advice provided by a licensed veterinary practitioner consulted in any particular situation. Octopus Publishing Group accepts no liability or responsibility for any consequences resulting from the use of or reliance upon the information contained herein. No dogs or puppies were harmed in the making of this book.

Contents

	Introduction to the second edition	6
	Preface	8
1	**The raw material**	10
2	**A puppy's view of the world**	20
3	**The new family**	30
4	**Developmental stages**	36
5	**Life with a new puppy**	40
6	**Socialization**	54
7	**Housetraining**	70
8	**Behaviour control and leadership**	78
9	**Toys and games**	92
10	**Preventing biting and aggression**	114
11	**Chewing**	130
12	**Handling and grooming**	134
13	**Good manners**	142
14	**Learning to be alone**	154
15	**Training your puppy**	156
16	**Adolescence and beyond**	196
	Appendix: Socialization programme	200
	Further reading	202
	Useful addresses	203
	Index	204
	Acknowledgements	208

Introduction to the second edition

A few weeks ago I received this email from Australia:

G'day. I purchased a copy of your book back in 1996 from a market in Melbourne, and read it from cover to cover before getting my dog (Australian Kelpie) in 1997. I have used the principles in your book on her since (she is now ten years old), she knows countless tricks and is a joy to be around. Thank you for publishing such a great book at a time when everyone else (well, in the area I'm from) used choke-collars!

Crystal Wemyss

I can't tell you how happy I am that the first edition of this book reached so many people around the world and resulted in a better life for their puppies. It has been the bestselling puppy book for the past 11 years, and I am so proud that many thousands of puppies and their owners will have benefited from what I was lucky enough to learn so early in my career.

When the book was first published, I used to hold my breath when owners came to tell me they had raised their puppy using what they had learned from its pages, half-expecting them to tell me that something terrible had gone wrong. However, this never happened and I continue to be amazed and delighted when they tell me how much the book helped them and how lovely their dog is.

Now, 11 years later, I get the chance to update the text and, using the principle of 'If it ain't broke, don't fix it', the changes are minor. Over the years I have learned what needs more emphasis, and I've

Puppies need to be kept occupied if they are to stay out of mischief and fit in easily with our busy lives.

Gwen Bailey, author and behaviourist, with another recruit for her education system that uses only rewards, play, fun and positive methods.

addressed in this edition some of the more common questions that new owners have. The biggest change comes in Chapter 8, 'Behaviour control and leadership', where I have tried to advise on the difficult question of how to get and keep control without using intimidation or punishment.

I think the success of this book has been due to it being based on positive methods. Punishment and aggression don't get us anywhere, whether it's with puppies, adult dogs or humans. Creating harmonious relationships based on love and trust is what this book is all about. Humans are by nature a very aggressive species and need no help in that department. Not that I'm against boundary-setting and enforcing standards of behaviour – and one of the main changes in this second edition is more information on how to do this well and appropriately. Having recently raised another puppy using these

methods, which has turned out to be happily well behaved and a delight to own, I am more convinced than ever that we are on the right track and that there is no place for any form of punishment in puppy education and training.

Dogs have always been the most important thing in my life, and I have dedicated most of my time to trying to get a better deal for them by educating their owners. Nothing has been more successful at doing this than this book, especially as it is read at a time in their lives when puppies (and owners) are still keen to learn and do the right thing. My hope is that this book will continue to be successful – perhaps even more so with its new, improved format and content. If we can all get it right at the start, fewer dog–owner relationships will founder or fail, and more will go on to be wonderful, long and happy. I hope yours will be one of them.

Introduction to the second edition **7**

Preface

Life for young puppies should be one long, happy adventure. Too often, it is a rather confusing time when humans expect too much of them, and puppies often get into trouble for breaking rules they did not know existed. In reality, dogs are not like Lassie. They do not automatically understand our every word and thought. They are a different species, with different capabilities and communication systems. They need our gentle assistance to guide them towards a better understanding of our ways and to help them learn what we want from them. Puppies, like children, require our love and protection. They need to be controlled just enough to make them nice to live with, but not so much that their spirit is stifled.

For more than 20 years I have worked with dogs that do not make it through the whole of their lives with the same owner. This is often because the first owners did not put enough effort into acquiring the knowledge needed to bring up a puppy correctly. If the partnership cannot be repaired, the loser is always the dog. Usually owners are not irresponsible or uncaring, but they often lack the necessary knowledge to do the job properly. It would take several dog lifetimes to get it right by trial and error, but it becomes much easier if we all learn from each other's mistakes and successes. This book is the result of what I have learned from my work with 'problem' dogs, but also from the knowledge

Ready and willing to learn. Young puppies can be moulded to our ways and it is up to us whether they turn out to be a pleasure or a pain when they are older.

Each puppy has his own temperament and personality. Appreciating what traits are inherent and guiding your puppy into good habits are essential if you are to produce a well-adjusted adult dog.

acquired through running puppy classes and seeing those dogs that I helped to educate develop into well-behaved, happy adults, as well as what I have learned from other professionals working in the field of animal behaviour.

At first glance, the amount of effort needed to rear a puppy correctly may seem daunting. However, this book is designed to help you bring up your puppy with the minimum of error and it covers all aspects thoroughly. It is unlikely that you will have to radically alter your natural ability to raise an infant, but this book will give you the extra knowledge needed to do the job well.

Planning in advance and getting things right first time around are quicker, in the long run, than having to sort out problems later. To make the best use of this book, put the ideas and suggestions into place *before* things begin to go wrong. In this way you will be able to avoid the problems that so many dog owners run into. This will make life easier, and better, for you and your puppy.

Key points

- This is a book solely about how to influence your puppy's behaviour and how to mould his future character. You will also need to find out how to care for your puppy's physical needs, but books on this subject are plentiful.

- I have referred constantly to 'he' rather than 'she', although there is no real reason for this other than to save writing 'he/she' or 'it' throughout. There is no difference between the worth of male and female dogs; both have qualities that can make them rewarding lifelong pets.

CHAPTER 1
The raw material

You may think that, on the inside, one puppy is much like another. In fact, by the time he gets to his new home, each puppy will be a unique individual ready to react in his own special way to his new environment, due to his own particular genetic make-up and the experiences he has already received in the litter. Whether you are still to make your choice of puppy or have acquired a puppy already, this chapter will help you understand how his background will influence his future behaviour and how you need to allow for these differences to ensure that he grows into a well-behaved and well-adjusted adult.

Breed groups

Dog breeds are classified by group, with each group having a set of general characteristics. Choose from the seven groups described in the table below.

Genes – a blueprint for behaviour

Our present-day breeds were created years ago by selecting useful traits for a particular type of work,

Collies are intelligent dogs that have been bred to work and need plenty of stimulating exercise.

BREED GROUPS	
Gundogs	Good-natured, sociable, biddable, energetic, boisterous, playful, medium to large in size
Pastoral	*Herders*: Reactive, sensitive, active, hard-working, energetic, playful, biddable, protective *Livestock guardians*: Usually large/giant, lazy, with thick coats and guarding traits
Terriers	Usually small, easily aroused and quick to resort to aggression, predatory, strong characters, excitable, can be noisy
Toys	Sweet, small, enjoy affection
Hounds	Amiable, independent, sociable, not that interested in toys, bred to track and chase, hard to control on walks
Working	Large, intelligent, protective; other traits depend on the work they were bred for
Utility	A group of dogs that do not fit into the other groups; individual characteristics will depend on what work the breed was raised to do

generation after generation, until each member of that particular breed has a predisposition towards a set of inherent behaviours typical of that breed. For example, if your puppy comes from a long-established line of Dachshunds, a breed originally developed to unearth badgers, he will be more prone to digging than, say, a Deerhound, which was bred to hunt deer. A Collie will be more likely to chase than a Cocker Spaniel. Dobermanns are more likely to be domineering than Dalmatians.

Inherited predispositions form the raw material that you will be working with and, in order to make the best of your particular puppy, it is helpful to know exactly what you have acquired. Information on what the ancestors of your puppy were bred to do can be found by reading books about your chosen breed, from the Internet, or from people knowledgeable about the breed. Think how these inherited traits will relate to the life he will lead with you. For example, do you live in a property where a lot of disturbance outside will trigger a guarding breed to bark too frequently? Do you live in an area where there are lots of exciting things to chase and there will

be consequences if he chases them? Does it matter if your dog digs in the garden?

Being aware of your puppy's inherited tendencies and desires will allow you to find positive ways to use up those energies so that they do not lead him into bad habits. For example, teaching a Collie to play chase-games with a ball may prevent him getting into trouble playing chase-games with joggers or cyclists. Rewarding a guarding breed for staying quiet and not reacting to outside noises when he is a puppy will ensure that he is not excessively noisy as an adult. Think hard about the temperament traits that your puppy has inherited, and find ways to channel those energies into good habits rather than allowing them to develop in an unacceptable way.

Finding out about inherited predispositions is less easy if you have a puppy of unknown parentage. If your puppy is a first cross from two pedigree dogs, it is fairly likely that he will inherit a mixture of the two sets of characteristics; hopefully the best parts of both. However, if you have a mongrel puppy, it is as well to acquire a knowledge of the various traits and to watch for their appearance in your own puppy.

Bloodhounds have been bred to use their nose to track and they have great energy and stamina.

CHARACTERISTICS OF THE 12 MOST POPULAR BREEDS

	BREED	ORIGINALLY BRED FOR	CHARACTER TRAITS
	Labrador	Retrieving fish	Like to carry objects and chew, energetic
	German Shepherd Dog	Herding and guarding	Chase and guarding, high work ethic
	Border Collie/ Working Sheepdog	Herding sheep	Need to chase, energetic, high work ethic
	Yorkshire Terrier	Killing rats	Can be predatory, easily aroused
	Golden Retriever	Retrieving game	Like to carry objects and chew, energetic
	Jack Russell Terrier	Killing rats and foxes	Predatory, easily aroused, energetic
	West Highland White Terrier	Killing rats and foxes	Predatory, easily aroused, energetic
	Cavalier King Charles Spaniel	Companion	Loving and affectionate
	Cocker Spaniel	Flushing game	Can be possessive
	English Springer Spaniel	Flushing game	Energetic, playful
	Boxer	Hunting bear and boar	Energetic and boisterous
	Staffordshire Bull Terrier	Fighting other dogs	Problematic with other dogs

SPECIAL CARE IN PET HOME	AMOUNT OF SOCIALIZATION NEEDED	STRENGTH OF WILL	ACTIVITY LEVEL
Teach and play retrieve with toys. Provide lots to chew on, especially during puppyhood. Provide plenty of mental and physical activity.	Average	Medium	High
Channel chase-tendencies into games with toys. Socialize and train well. Provide plenty of mental and physical activity.	Lots	Medium to strong	High
Channel chase-tendencies into games with toys. Provide plenty of mental and physical activity.	Lots	Medium	Very high
Care needed with small pets. Not usually a problem, due to their small size.	Average	Medium	Low
Teach and play retrieve with toys. Provide lots of chews, especially during puppyhood. Provide plenty of mental and physical activity.	Average	Medium	Medium
Care needed with small pets. Good socialization and handling required. Provide plenty of mental and physical activity.	Average	Strong	High
Care needed with small pets. Good socialization and handling required. Provide plenty of mental and physical activity.	Average	Strong	High
No special care needed.	Average	Weak	Low
Teach and play retrieve with toys.	Average	Strong	Medium
Teach and play retrieve with toys. Provide plenty of mental and physical activity.	Average	Average	Very high
Provide plenty of mental and physical activity. Play with toys. Train.	Average	Average	Very high
Ensure adequate socialization with other dogs.	Average	Strong	Medium

Which breed is for you?

Many people choose a dog simply because they like the way he looks, rather than considering how he is likely to behave. This is not the best basis on which to choose a companion that will probably be living with you and your family for, possibly, the next 15 or more years. Finding the puppy with the genetic make-up that suits you and your family will make it more likely that you will succeed in raising a well-adjusted pet dog. It is, therefore, worth giving this part of the equation considerable thought beforehand. If you have already acquired your puppy and have realized that you may not have chosen wisely, don't despair. It is possible to turn a puppy that does not have the best genetic make-up for you into a perfect dog; but you may have to be a little more accommodating, work a bit harder and accumulate a greater knowledge in order to do so.

Working strain or show stock?

During the early development of the breeds, dogs were selected for how they acted rather than for how they looked. Times have changed, however, and now most pet dogs come from show stock, where the predominant requirement is to produce a dog that is 'typical' of the breed in appearance only.

Since physical appearance alone is tested in the show ring, only the more caring breeders are

Choosing a puppy that will be patient, affectionate and tolerant is important if you have children. Labrador retrievers are a popular choice for people with families.

Energetic Springer Spaniels are lively and affectionate and make great pets providing someone is willing to exercise them well.

concerned about the temperament of individuals in their particular breeding line and take steps to ensure that only dogs with sound characters are used for breeding. Even then, the temptation to use a stud dog with a less-than-perfect temperament, which, nonetheless, has the best conformation to suit their brood bitch and hence is more likely to produce champions in the show ring, may be too much for some breeders.

Some dogs are still bred for their working ability. Sheepdogs are a prime example. The best puppies from each litter will be kept on, to be trained and worked. The surplus will find their way into pet homes. Before you take on a dog with this sort of background, consider whether you want a dog that is capable of running 32–48 km (20–30 miles) a day and has the stamina to keep going all day every day. Kept in an ordinary pet home, where the owners have to go out to work for most of the day, working dogs such as these can drive their humans mad, unless

they become demented themselves in the process. So if the breeder proudly shows you photographs of the parents at the final of a Sheepdog trial, or winning an award for Best Police Dog of the Year, or competing in the Iditarod sled-dog race across Alaska, you would be wise to consider carefully whether you want a pet dog with these inherited abilities. In addition, working dogs are often kennelled, so that puppies do not always get the early socialization they need for life in a pet home.

Unfortunately, few people breed 'pet' dogs in Britain. Dogs are usually bred for their beauty, working ability or by accident. Only by finding out as much as you can about the various forms of ancestry on offer can you make an informed choice as to which line of breeding will be right for you. If you already have a puppy, finding out this type of information will help you understand which character traits he may develop as he matures, so that you can be ready to help him give them acceptable expression.

A good breeder will keep puppies in comfortable and hygienic conditions and ensure that they are well grown and well socialized before offering them for sale.

Where your puppy comes from

The source of your puppy is very important. As we shall see later (see page 56), the process of socialization should already be under way by the time puppies are ready to leave their mother, and how thoroughly this has been done will make an enormous difference to the way your puppy eventually turns out. If this has not been done well by the breeder, you will need to work very hard while your puppy is still young, to ensure he grows up well adjusted and unafraid.

The best possible start for a puppy is to be born into a busy, lively household where he can experience all the sights and sounds that he needs to become familiar with. If he is handled (carefully) by children and adults every day, played with by visitors and has met other friendly dogs, by the time he gets to you he will already be on the way to being well balanced and confident in all situations. If, on top of all of this, the litter was planned, and care has been

taken to breed from parents with sound, friendly temperaments, you have the best possible recipe for future success.

Puppies from other sources can turn out well, provided you are careful in your choice. Breeders who keep their dogs in kennels, but take time to bring them into the house and socialize them, are rare, but can produce good puppies; as can some rescue centres that keep their puppies in a place where they get plenty of human contact and different experiences from an early age. Buying a puppy that you know has had plenty of pleasant experiences in a varied environment will set you off on the right foot.

Beware of buying a puppy from sources where little care has been taken – for example, a farm where the puppies are kept outside and have never been taken into the home; a show kennel where no effort has been made to handle or socialize the pups; a pet shop where there is no way of knowing how the puppies have been raised; or a puppy-farm outlet where large numbers of puppies of different breeds are brought together, just to offer the customer a wide choice.

Puppy-farmed dogs, bred *en masse* in unsuitable conditions, specifically for the pet trade, are well known for having temperamental and physical disorders. It is unlikely that care will have been taken to select the parents; often any dog that looks vaguely like the breed required will do, and the early stress and trauma of being born in such circumstances and then transported long distances at a very early age take their toll. Some peculiar behaviour problems and temperament disturbances have been seen in dogs bred in this way.

To make sure you are not buying such a puppy, insist on seeing the puppy with his mother, and be very suspicious if the breeder makes excuses as to why this is not possible. Go back for another visit,

if necessary. Never take on a puppy that the breeder or supplier offers to deliver to your home, or if he offers to meet you halfway 'to save you a long journey'. Beware of advertisements that offer puppies of several breeds.

Choosing a puppy

Select a puppy that, at the age of six to eight weeks, will approach you readily with a confident posture and a happily wagging tail. Well-socialized puppies are pleased to greet new people in a calm, friendly manner and should be content when picked up and gently restrained. Avoid puppies that flatten as they approach or that try to avoid you. If you have already bought a puppy that is concerned about human contact, you will need to work very hard to overcome this fear and shyness while your puppy is still young (see page 68).

Golden Retrievers usually make lovely playful pets. Try to choose a puppy that has no possessive aggression in the breeding lines.

Seeing the mother with the puppies gives a good indication of your puppy's future temperament. Nervous traits are usually inherited, and it is also likely that puppies that experience their mother's fear and aggression towards strangers at an early age learn this behaviour, and will show it later as they mature. Often, when I have seen people whose puppies have a nervous-aggression problem and ask if they saw their dog's mother, they say, 'Oh yes, but you couldn't get near her' or, worse, 'Yes, but she nipped me.' One couple, whose dog was very aggressive towards strangers, and who had bought the puppy knowing that his parents were guard dogs, commented that they wanted their dog to be a guard too, but not to bite people! Breeders may tell you that the mother cannot be seen, because she has become aggressive to strangers since having her puppies;

this means that she had an underlying aggression problem that has been revealed with the stresses of giving birth and the extra visitors, so it is not wise to take one of her puppies.

Lastly, beware of the many unscrupulous breeders and dealers who use puppies as a way of making money. I know of one man who breeds German Shepherd Dogs that have highly unreliable temperaments and sells them with the assurance that he will take them back, should anything go wrong. He usually gets them back again at one year old, because the owners are unable to handle them, and then he sells them for a second time as guard dogs.

When to take your puppy home

Opinions vary on the best age for a puppy to leave his mother and litter-mates to go to his human family.

A puppy like this one, who has had good socialization with the breeder and who is happy and confident, will give you the best chance for success.

While he is with the breeder, a puppy needs plenty of different types of positive experience so that he feels confident and at ease going into a new family.

The advantages of staying in the litter have to be weighed against the advantages of being with the new family.

The longer the puppy stays with the mother and his litter-mates, the more he will learn about canine communication systems and the better he will be able to cope with encounters with other dogs later in life. Puppies leaving the litter too soon – as in the case of orphaned puppies that need to be hand-reared – miss out on play with other puppies. They may be unable to deal adequately with encounters with other dogs, which may lead to problems with other dogs when they mature. They also miss out on vital education from the mother, as she cares for and weans her puppies. Puppies leaving the litter too soon can be difficult and aggressive when they cannot get their own way, because they have never learned to deal with feelings of frustration, as happens naturally during the weaning process.

However, the longer a puppy stays in the litter, the less chance he has to learn human ways. If a puppy stays in a litter too long – as in the case of puppies that are 'run-on' by breeders until about six months of age to see if they develop sufficiently well for show purposes – he will be far less competent in encounters with humans, making him a very poor pet dog. Such dogs often enjoy the company of other dogs more than humans, are difficult to communicate and play with, and may be shy and more prone to showing nervous aggression towards strangers.

An added complication is that puppies of smaller breeds tend to mature faster than puppies of larger breeds. For example, a small terrier puppy may be at the same stage of development at six weeks as a puppy from a giant breed is at eight weeks.

So the decision about when to take a puppy home has to be a compromise. Since it is more important for pet dogs to be able to interact well with people rather than other dogs, this should be given more consideration. If the puppy is being well socialized with adults and children and getting plenty of new experiences every day, it may be best to leave him with the breeder until he is eight weeks old. If he is not, then six weeks may be a better age for him to go to his new home, where time is better spent learning to be part of a human family. Do not take a puppy that is older than eight weeks, unless you know for certain that he has already been well socialized with humans and has had many varied experiences and plenty of individual attention.

CHAPTER 2
A puppy's view of the world

Ancestors of our domestic dogs, the wolves, evolved long ago to be cooperative hunters of large prey. Their brains and bodies developed to serve this purpose and to allow them to detect, chase and kill large prey by cooperating and coexisting with each other in packs. Consequently dogs have different motivations, senses and abilities to us and their view of the world is very different from ours. To bring up a puppy successfully, it is helpful to be able to look at things from their point of view.

Living in a world of scent

Sight is our primary sense and we learn most about our world by seeing with our eyes. In dogs, the sense of smell is much more important, and much of the information they gather from their environment goes

Puppies are small relative to us and hands reaching down to grab them from above may appear frightening until they get used to them.

Puppies live in a world where scent is very important and gather a great deal of information about their surroundings through their noses.

in through their nose. Watch a dog and owner as they enter a new room. The human will use his eyes to find out what goes on there, whereas the dog will go sniffing round to discover what he needs to know.

Dogs can detect odours in a way that we find hard to contemplate. Trained dogs can easily follow the route taken by a person who passed by hours (sometimes days) earlier, leaving no visible signs, or they can sniff out minute amounts of drugs or explosives through layers of packaging and containers.

Their sense of smell is known to be at least a hundred times better than our own, and maybe even more. The area inside the dog's nose that detects scent is about 14 times larger than ours, and the part of their brain that processes information

is proportionately larger than ours and better developed. Consequently dogs are not only better able to detect smells than we are, but they are more interested in them, too.

In the wild, this would have been extremely useful, not only for the detection of prey animals, but also for the maintenance of social groups and the defence of territory. Being able to tell who your friends and enemies are from a distance is very handy. Knowing the sex, state of health, age and reproductive state from one sniff saves a lot of questions!

This amazing sense of smell is a feature that has been handed down to our own pet dogs, and it helps to explain why they are so fascinated by scents, and why they go to great lengths to gather information

A puppy's view of the world **21**

Dogs can hear sounds in the ultrasonic range, which explains why they can learn to respond to a 'silent' dog whistle that we cannot hear.

through their nose. By sniffing every lamp-post or putting their nose into all the wrong places when investigating new people, dogs are gaining information about their environment that may be useful to them later, in much the same way that you and I get clues about our environment using our eyes. Puppies will often recognise you instantly by smell but take longer to learn to distinguish you by sight.

Amazing hearing

Dogs are more sensitive to sound than we are. Sounds that can only just be heard by us at a certain distance can be detected by dogs that are four times as far away. So there is no need to shout!

They can also hear a higher range of frequencies, which means that they hear sounds in the ultrasonic range that we cannot detect. In the wild, this enables

Hearing range

As well as being able to hear sounds from further away, dogs can also hear sounds of a high frequency, allowing them to detect noises of prey animals. A dog's hearing range extends from 40–60,000 Hz and a human's from 20–20,000 Hz.

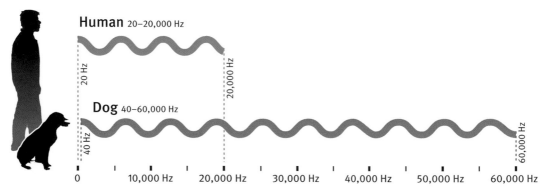

Human 20–20,000 Hz

20 Hz

20,000 Hz

Dog 40–60,000 Hz

40 Hz

60,000 Hz

0   10,000 Hz   20,000 Hz   30,000 Hz   40,000 Hz   50,000 Hz   60,000 Hz

How dogs see the world

Dogs see differently to us. They have a wider field of vision, perceive detail and texture less clearly and do not see red or green. They see better in dim conditions and are more sensitive to movement, allowing them to detect prey easily.

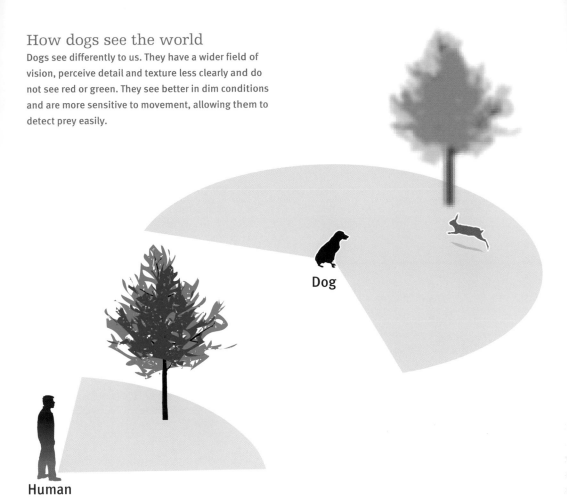

Dog

Human

them to locate small prey, such as rodents, which communicate in squeaks at a very high frequency. This ability explains why dogs can respond to 'silent' dog whistles when we hear nothing.

Some breeds of dog, such as Collies, which have been bred to enable them to hear a shepherd a long distance away, have more sensitive hearing than others. This explains why they develop phobias about thunder and gunfire so easily. If these noises sound loud to us, imagine what they must sound like to a puppy with sensitive ears.

Less-detailed vision

Dogs see less well than we do. Intricate things that we can see in sharp detail appear blurred to them, and they recognize objects by shape and form, rather than by detail and texture. They are not colour-blind, but they cannot see red and green, so their vision consists of yellows, blues and shades of grey. This makes it more difficult for them to perceive certain objects that we can see clearly, such as a red ball on green grass.

Dogs can see better at night and in dim conditions than humans can because they have a reflective layer at the back of their eyes, which traps any light that enters and enables them to make more use of it. This is why they can run off at top speed into the darkness on winter walks without crashing into trees and fences (and why their eyes 'shine' when caught in the beam of a car's headlights).

Dogs are also much more sensitive to movement than we are, especially any that occurs at ground

Visible colour spectrum

As these visible spectrums show, dogs cannot see red and green so their vision consists of blues, yellows and shades of grey. This is important to remember when we are playing with them with coloured toys.

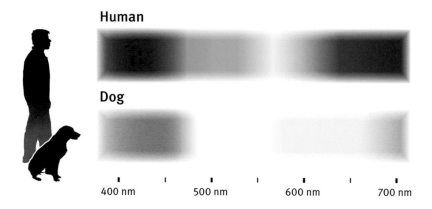

Human

Dog

400 nm   500 nm   600 nm   700 nm

What we see (*top*) and what dogs see (*bottom*). We can perceive red and green objects clearly whereas our dogs will only see them as shades of yellow and blue. The world through a dog's eyes is much less colourful.

level. We are able to see stationary and moving objects equally well, whereas dogs are much more likely to see objects that are moving and to ignore stationary ones. This sensitivity means they can detect the slightest movement of our bodies, and allows them to anticipate our actions before we have deliberately moved.

Body language and the spoken word

Dogs communicate with each other using body language. This involves posturing with tail, ears, body position, eye contact and facial expression. A lot of information can be passed between two dogs in this way, and it is their substitute for the spoken language that we rely on.

In the wild, dogs have little use for vocalization and so they find our words relatively difficult to learn. This explains why they learn hand signals more readily than spoken voice-cues.

Since dogs have such a different language system from that of humans, difficulties of communication frequently occur between the two species when they interact. Dogs often misinterpret our intentions and humans often misread their dogs, which leads to all sorts of problems. To overcome this, you will need to learn to read your puppy's body language so that you can find out if he is scared, unhappy, tired or joyful. You also need carefully and patiently to teach your

Dogs learn to read our body language easily because they communicate with each other visually. They will learn gestures and signals much more quickly than they learn spoken words.

puppy to respond to each spoken voice-cue that you want to use (see page 156).

You can also improve communication with your puppy by using highly obvious body postures and signals. Puppies will learn these much more easily than the spoken word and training will be faster if you use them (see page 168). Puppies understand some of our body postures without training, for example, towering over your puppy with a stiffened posture, staring eyes and a tight mouth signals an intended threat.

Your puppy will watch you more than you expect, and will read your body signals rather than listen to what you are saying. As he becomes more experienced, he will quickly learn to interpret your mood and will know what you are feeling, even without you saying anything. He will also learn to respond to the many things you do that he would not understand naturally such as crouching down and sending 'open' body signals as welcome sign. Help him out at first by being consistent with your signals and making it obvious what you intend him to do.

A puppy's view of the world **25**

Body language What's he saying?

▷ **Look I'm really big so don't bite me!**
This puppy is aroused and worried. His hackles are raised all along his back and he has put up his tail to make himself look bigger. The other puppy sits down and puts his ears back because he is worried too.

△ **You are so far beneath me** The adult dog is stiff with a raised tail and averted eyes, indicating that he doesn't want to interact with this puppy. The puppy is confused by this response to his playful overtures and engages in a little displacement, scratching to make himself feel better.

I'm scared Ears back, tense face and body, and tail between legs all indicate that this puppy is worried. The forepaws slope backwards if the weight is transferred to the back legs, ready to run away if necessary. Puppies showing these signs need help and support to overcome their fears.

Just relaxing This puppy's body is relaxed and soft. The ears are forward and the tail is gently swinging. Contrast this with the tense bodies and faces of the puppies in the other photos.

Trying to feel better Puppies often lick their noses or yawn when they feel worried or under pressure. This puppy looks tense, but not yet scared enough for his ears to be drawn back. If he continues to be worried he will get up, as he is vulnerable in the 'down' position.

Let's play! Both of these dogs feel comfortable with each other and want to play. They invite each other to have a game by play-bowing with elbows on the floor and bottoms in the air. The Labrador is a little less at ease than the terrier and gives her lack of confidence away with a tense body and stiff tail.

What's going on? The Labrador puppy has relaxed now and is inquisitive. He shows his interest by standing over the terrier puppy, who continues to worry. The terrier's lowered head, which is turned away from the Labrador with ears back, shows his concern.

Direct eye contact and a smile that shows the teeth is threatening in dogs and causes this puppy to look away despite the treat that is being held under the chin.

Facial awareness

Humans' actions, such as smiling and staring, can be misinterpreted. Showing the teeth in humans is a sign of friendship, whereas in a dog pack it is an obvious signal that the pack member has the capability to bite and is likely to do so if the provocation continues.

Dogs often use direct eye contact to threaten each other. An adult dog may caution a misbehaving puppy by staring at him; or a high-ranking dog, using piercing eye contact, will indicate to a subordinate that it should not proceed further with what it is doing. Humans sometimes do this with a glare, but most human eye contact is friendly and humans often stare lovingly at dogs with eyes wide open.

Puppies being brought up in a human family need to learn the difference between the two types of eye contact. They need to know that another dog staring at them across the park may mean business, whereas a new human friend who is staring at them is being friendly. This should happen quite naturally during the socialization process, although some shy puppies may need extra help to learn this from lots of friendly contact with the owner.

Mouths instead of hands

It may seem obvious, but dogs do not have hands, so they tend to use their mouths to pick up and explore objects instead. The whiskers around a dog's face help with the process of exploration. The bases of these whiskers feed information directly to the brain and are very sensitive to touch. Until they have got used to it, puppies often dislike having their whiskers touched by humans, and will move their faces away if hands are placed too near them.

CHAPTER 3
The new family

The household that a puppy is brought up in will have a significant impact on his future character. Who lives in the household – whether they are old or young, aggressive or timid, happy or miserable – will leave an impression on him. Whether the people in the house have previously owned dogs or not, or whether there are children, dogs or other pets in the family, will also have an influence.

Reflections of ourselves

If you look at a group of ten-month-old puppies and their owners, it is relatively easy to tell which puppy belongs to which owner. This is because puppies tend to directly reflect their owner's characters as they mature, possibly because they share the same emotional experiences. Happy, loving people, for example, tend to have happy, outgoing puppies, whereas miserable, boring people tend to have puppies that are withdrawn and disinterested.

Any children in the family tend to be reflections of their parents, too. If, during puppy classes, the children sit quietly, politely listening to what you have to say, the chances are their puppy will be well behaved. If the children are not well controlled by their parents and keep interrupting, despite being told not to, there is a strong possibility that their puppy will be over-boisterous and wilful. This is so because people tend to bring up their children and puppies in the same way they were brought up themselves. The influences that shaped their characters are similar to those that will be shaping their children's and puppy's characters, so it is likely that they will all have similar ways of behaving.

Think about how you were brought up by your parents. If you were punished a lot as a child and over-controlled, consider what effect this will have on the puppy you are about to raise. If you were brought up in a way that made you extrovert and outgoing, how will this affect your puppy? Were your parents always shouting at you, in a vain attempt to make you obey, or were they quietly in control?

Take a look at your own family and any children that you may have. Do you have the kind of temperament that you would like to see in an adult

Raising a puppy and children together can be good for both of them providing both learn acceptable behaviour and good manners.

Puppies will take on many of our temperament traits as they adapt to life with us, becoming a reflection of ourselves and our children.

dog? How quickly do you get angry? How placid are you? Are you outgoing and lively, or are you more calm and introverted? If you add all of the family's characteristics to those of the breed of the puppy you have chosen, it will give you a good indication of what your adult dog will be like. If your prediction of the future characteristics of your puppy is not the ideal you would like to see, consider how to change the way you bring him up so that he develops differently.

Once you are aware that you have a choice about how you bring up your puppy – and that you do not have to follow the example set by your parents – it becomes much easier to decide on your method of upbringing. Have a look at dogs belonging to other people, if necessary, and see if you like their characters. If so, find out how they were raised.

Discuss the method of upbringing with the rest of your family so that you are all in agreement, and you should then have set the scene for producing an adult dog with a temperament to suit you all.

Living alone

Situations where there is only one person and one dog in the household tend to produce intense relationships. So much affection, time and effort are expended on the relationship that other people can be excluded. This often produces a dog that is intolerant of outsiders, particularly those coming onto his territory.

In addition to the isolation from the outside world, single owners will often give the puppy privileges that family owners will be too busy to offer. As a result, the puppy may grow up to think that he is equal in status with his owner and this, coupled with a lack of socialization, can lead to unwanted aggressive protection of the owner.

If you live alone, you will need to work extra hard at socializing your puppy. You will also need to guard against a relationship that is too inter-dependent and and avoid giving him too many privileges.

Children in the family

Families with children tend to be lively and busy, which is good for socialization purposes. The negative side is that often too much is happening to pay close attention to the puppy's education and the owners are swept along on the tide of events, suddenly noticing one day that the puppy days are over and they now have a boisterous, untrained adult dog to deal with.

Sometimes parents are too busy to have time for exercising, playing and educating the puppy, which is often left to the children. If left to their own devices, children (especially young ones) may unintentionally teach a puppy bad manners. Leave a young puppy to play unsupervised throughout his puppyhood with several young children, and you will end up with an adult dog that has learned to chase, jump up and nip at their legs or arms. Being stimulated by movement

Getting to know each other and adapting to each other's ways is good for both children and puppies.

and a desire to join in their games, this is natural behaviour for a puppy.

A puppy that learns to have fun by excitedly chasing children and nipping at their ankles will not see the harm in doing this when he is an adult. Children may think it is funny while the puppy is still small and may encourage it, but will not think it so clever when the dog is fully grown. Worse still, children in the park who do not know your dog may not realize that he is intent on play when he bounds towards them. No matter how friendly your dog was being, it could scare a child enough for the authorities to believe your dog was dangerously out of control, which might lead to prosecution.

If you have children in the family, you need to ensure that they teach your puppy only good behaviour. Instruct the children what to do when the puppy jumps up or pulls at their clothes; teach them how to let the puppy know what they want and how

to play acceptable games. Discreetly supervising all their activities will prevent either side from learning or doing the wrong thing.

Remember that toddlers are only little themselves and will pinch, pull and throw things that may hurt the puppy. Puppies, with their needle-sharp teeth, can also hurt, and you will need to be there to intervene on behalf of either or both parties, if necessary. Older children can tease or be intentionally cruel just because they are at that age when they are finding out about their world. Teenagers are generally much too interested in their own lives to show more than a passing interest in a puppy, and it is probably not wise to rely on them to supply the puppy with what he needs. However, all children are capable of showing great love and affection to a puppy and often make better playmates than adults if shown the right direction. By giving the situation careful thought, bringing up a puppy with your children can be – and should be – a pleasure and an education for them both.

Couples as owners

Couples generally make good puppy parents. If you are planning to have children in future, make sure you socialize your puppy well with babies and toddlers.

Another dog in the household

Many owners acquire a new puppy as company for their existing dog, especially if they are away at work for long periods each day. Unless they take steps to prevent it, this can lead to the puppy learning to relate more to the other dog than he does to his owners, and they may find themselves with an adult dog whose behaviour is a lot less than perfect.

A puppy on his own in a family without another dog will have to learn human ways, because humans are his only source of companionship. However, a puppy that has another dog in the house already has a friend that can speak the same language and play the same games. He will have no urgent need to learn the ways of humans.

This is similar to the security we feel if we travel to foreign country with a friend who can speak our own language. They provide the social contact we need, so it is not so important to risk trying to speak to strangers with whom it is difficult to communicate. This often leads to a situation where your puppy will prefer to be with your other dog rather than with you.

As the puppy matures into an adult and develops more confidence, he will be less willing to do as you say and more difficult to live with as a result. All sorts of control issues can arise because the relationship between you is not close. For example, he may prefer to play with other dogs rather than you in the park, and so may refuse to come when you call him or may actively run away to find other dogs to play with.

If you have an older dog in the family, or have two puppies from the same litter, or you regularly (nearly every day) meet up with another dog that your puppy plays with, you will need to ensure the relationship between your puppy and humans develops properly. It is important to spend more time playing with him than he spends playing with other dogs. He can still play with other dogs and puppies – this is important for socialization – but this time needs to be limited.

Raising two puppies at the same time is not recommended as much time and effort are needed to give both the attention they deserve.

The quality of play with you will not be as good as it is with other dogs, so aim to spend at least three times as much time playing with your puppy as he spends with other dogs each day. If he spends five minutes with the other dog, you need to play with him for at least 15 minutes, split up into five-minute sessions.

In order to achieve this, you will need to stop them playing together when you are present, and to separate him from your other dog if they are left alone together, either during the day while at work or at night. It is best to have a mesh partition between them (in some cases a stair gate works well, as long as the older dog does not jump over it). This enables them to keep each other company, but they cannot play together unless you are there to supervise.

When you want to spend time with your puppy, first confine him while you have a good game with your older dog. Then put the older dog out of the room while you give your attention to the puppy. Otherwise your older dog will try to join in, the puppy will try to play with the older dog, you will get frustrated because your puppy doesn't want to play with you

and, because you are cross, your puppy will try harder to play with the other dog and keep away from you!

Keeping control of all the contact that your puppy has with other dogs until he has reached maturity, and ensuring that he spends more time with humans than with other dogs, will help him to grow up as a human-oriented dog. He will be much easier to train and live with than a dog that likes other dogs best. You will have formed a strong bond between the two of you, and your dog will relate better to you and to all other humans as a result.

Two from the same litter?

Unscrupulous breeders often encourage new owners to take two puppies from the same litter, in order to sell more puppies. This is never a good idea. The bond that develops between them – already strong because they are siblings – will be stronger than any bond they have with their owners if they are allowed to play and have continual access to each other.

In order to prevent this, you need to find as much time for both puppies individually, for play, training

Litter-mates will have a very strong bond, making it difficult for you to form good relationships with them.

The new puppy and existing dog can become good friends if they are introduced carefully.

and general companionship, as most owners would give to one. For this reason, unless you have a great deal of time and energy, it is inadvisable to try to bring up two puppies of the same age simultaneously. It is better to raise one puppy until he reaches social maturity at about 18 months to two years old, and then get a second.

Other pets

Having other pets in the household can help to socialize your puppy with other species and give you a chance to teach control during exciting situations. If a young puppy is introduced to a pet of another species while he is still very young, he will usually accept it as just another member of the family. And as an adult, your dog, having been brought up with other pets, is more likely to tolerate the introduction of similar pets as you acquire them.

However, the instinct to chase, catch and kill small, fast-moving animals can sometimes overcome dogs, and no dog should be trusted alone with small, easily damaged prey species.

Dogs alone and together

- It is important that your puppy learns to be left alone completely, without the company of the other dog (see page 154). It is also a good idea to take your puppy out without your older dog. You can then devote more attention to him, and he will learn to be confident when out alone and not to rely on the other dog for support.

- Once they know each other, do not interfere if your older dog decides to discipline your puppy by growling, snapping or holding him down, unless you fear for your puppy's safety. Sometimes your adult dog will need to tell your puppy to stop and lie quietly. By interfering, you will be altering the natural balance that otherwise exists between them.

CHAPTER 4
Developmental stages

All puppies follow the same pattern of development, passing through the same stages from infancy to maturity. It is important to know these developmental stages and to realize what responses your puppy is capable of at any particular time in his life. This is so that you do not expect too much from him, or miss out on making the most of an opportunity at a critical stage in his development.

The speed at which puppies progress varies a little: some individuals pass through some stages quickly, while others take longer than expected. Generally, puppies of smaller breeds tend to develop more rapidly, often reaching maturity before one year old, whereas puppies from larger breeds can take longer, some taking 18 months before maturing fully. What follows is an average timescale to which most puppies conform.

Newborn period, 0–2 weeks

During this short phase the puppy mostly sleeps and suckles. He can crawl and will try to find warmth if cold. He needs the mother to stimulate urination and defecation, which she does by licking the genital area. The eyes open at around 10–14 days, but vision is poor for the first few weeks. Gentle handling is all that is needed at this stage.

Transitional period, 2–3 weeks

The teeth begin to appear. The puppy learns to walk and lap liquids. The ears open towards the end of the third week and the sense of smell begins to operate. The puppy develops the ability to urinate and defecate by himself.

What to do: The responsibility for what happens to the puppies during this time lies with the breeder. It has been found that puppies subjected to mild stress at this point are better able to cope with other stresses later on. Picking up each puppy every day, looking at them and perhaps weighing them constitutes mild stress, and conscientious breeders will do this. Each puppy should be picked up once a day and gently held in different positions.

Socialization period, 3–12 weeks

During this critical period appropriate experience with humans, other dogs and the environment is essential if the puppy is to develop into a successful pet dog. This period can be divided into three stages.

First stage: weeks 3–5

At three to four weeks, the puppy's sight, hearing and sense of smell are becoming more efficient. He

2 weeks

3 weeks

5 weeks

begins to eat solid food, bark, wag his tail and play-bite other pups. He will attempt to leave the sleeping area to urinate after waking at this age.

At four to five weeks, he paws, bares his teeth, growls, chases and plays prey-killing (head-shaking) games. He begins to carry objects in his mouth. The puppies begin to learn to inhibit their bite during play with their litter-mates. Mothers will begin to prevent her puppies from feeding at will.

Second stage: weeks 5–8
Facial and ear expressiveness are seen now, and weaning begins. Puppies acquire the full use of their eyes and ears and become more coordinated. Participation in group activities with others in the litter is seen; games are played between litter-mates. The seventh week is the ideal time for puppies to go to their new home. At the end of this stage puppies begin to be more cautious, but they are still curious and will investigate anything.

What to do: During weeks three to six, the puppy will still be with the breeder. There should be a clear distinction between sleeping and play areas, so that puppies can leave the nest to go to the toilet. Puppies kept in too small a space, in one where there is no distinction between the areas, or where they cannot get out of the nest box unaided, do not have the chance to practise appropriate toileting behaviour. Puppies that are raised in this way can be difficult to housetrain because they lose the ability to differentiate between nest and toilet area.

The puppies' rate of mental development will now depend on the complexity of their environment. An assortment of different objects (such as cardboard boxes, a large piece of hard leather, toys or an old glove) should be placed in the pen with the puppies to provide this complexity. It is also helpful to have a few low, wide steps for the puppies to practise on. These will help them to negotiate stairs later.

During this time a variety of noises, sounds and different floor surfaces should also be encountered. This will probably happen naturally for puppies raised in a home environment.

Puppies should be isolated occasionally from their litter-mates and their mother, and there should be plenty of contact with humans, both adults and children. This is particularly important in the final week before the puppy goes to the new home, and at least five minutes of individual attention should be given to each puppy every day, no matter how many puppies there are in the litter.

During weeks six to eight, the puppy should be settling in with his new family. Socialization will continue naturally as the puppy experiences the novelty of his new home. Housetraining should begin as soon as the puppy is brought home.

Third stage: weeks 8–12
The puppy now explores away from the nest. He is still very dependent on his attachment figures, and has a strong desire to please and a need for social contact. Puppies of this age become increasingly afraid of things they have not encountered before.

8 weeks

12 weeks

What to do: Your puppy should be experiencing and enjoying a wide variety of situations and environments (see page 54). This is a critical time for socializing and you should make the most of it, before the fear of novelty begins to outweigh the interest in new activities. Play is extremely important, and your puppy will need to learn to play human games and to reduce and inhibit his play-biting (see page 96).

Juvenile period, 3–6 months

The puppy is still dependent on his owner. In the early stages of this period, puppies are usually eager to please and will do whatever they believe their owners want them to do.

Environmental awareness is increasing now. Puppies begin to explore further afield, but always stay within range of the security of their owner or familiar territory. Chewing and mouthing behaviours are common, to facilitate teething and aid environmental exploration.

What to do: Training and good manners should be developed as the puppy becomes better able to concentrate and learn. The willingness-to-please of

this age should be used to the full, as training will be more difficult later when the puppy becomes more independent. Games can become more advanced and can be used as a reward for training.

Your puppy should experience many different environments and circumstances. Socialization should continue and be developed as the puppy learns to cope with new situations.

Adolescence, 6 months–1 year/ 18 months

Puppies become much more independent at this stage. They reach sexual maturity; females come into season, with associated behavioural changes, and males experience dramatic fluctuations in male hormone levels. Chewing behaviours continue to be a priority. Territorial behaviour starts to appear as puppies become more mature.

What to do: This is possibly the most difficult period to live through, and is a time when many people give up their dogs for rehoming. If you have laid down solid foundations of good behaviour up to this point, adolescence will be less wearing. However, it will be a

6 months

1 year

time when you may wonder what has gone wrong. Try to remember that it does not last for ever! (See page 196 for help.)

Maturity, 1 year/18 months

Your dog will now be physically mature, although he will have some filling out to do and his character is still developing. Young adults have the confidence of youth, but are inexperienced.

What to do: Care is needed to restrain the enthusiasm of his interest in the world while he continues to learn from new experiences. Keep up the training and use physical control if necessary to keep him out of any dangerous situations he encounters.

Social maturity, 18 months/ 3 years onwards

Your dog will now be both physically and socially mature. His character is formed, and he will have tempered his earlier enthusiasm for life with the experience that comes with maturity. He will be an experienced pet dog that knows his role in life and is ready to respond to anything that is expected of him.

What to do: You can relax now – your job is done. Refinement and continuation of training are needed, but you should now be able to take it easy and enjoy the first of many happy years with your well-balanced, well-trained, sociable friend.

3 years

18 months

CHAPTER 5
Life with a new puppy

First impressions really do count, so take care when introducing a puppy to your children, dogs and other pets. Great excitement is usually generated when a new puppy arrives home, particularly where children are concerned. Try to diffuse this as much as possible and to keep introductions low-key, quickly distracting the other occupants of the house as soon as the first meeting is over to allow the puppy to explore and the excitement to subside.

Introducing children

Ask children to sit down when the puppy is brought in, and to let the puppy approach them. Give them small treats to feed on the flat of their hands so that the puppy gets a good first experience. Buying children a new game or toy at the same time as you bring the puppy home may help to distract their attention and give the puppy some much-needed space to find his feet.

Teaching children to use a calm, gentle approach is essential when a new puppy joins the household, especially if he is a little shy.

This puppy is a little daunted by his first meeting with his large friend. The adult's relaxed approach shows that all will go well once they get to know each other.

It is helpful, especially if you have young children, to make a rule that the puppy is not to be picked up. This will enable him to learn that children are friendly and nice to be with. He cannot learn this if he is whisked off his feet each time he meets them. Also make it a rule that your puppy is not to be disturbed when sleeping as this will make your puppy irritable very quickly.

Introducing other dogs

Aggression towards puppies usually occurs because an older dog is frightened by the puppy approaching and running round its legs. If the older dog is likely to object to this, use a lead and small harness to prevent your puppy getting too close too quickly, and keep meetings brief and pleasant until the two dogs start to accept each other.

If you can find a suitable place, the first meeting is best done away from home, while out on a walk. Exercise your dog well on the way to collect your puppy, and keep them separate in the car. On the way home take them both out to a new area, somewhere your dog has not been before and away from areas where other dogs have been (your puppy will not be fully protected by vaccination yet). Keep walking slowly and try not to interfere too much, but keep hold of one for a while if the other looks unhappy with the situation. The distractions of the interesting new sights and smells will take away the intensity of the meeting, and both dogs are less likely to be worried about the situation than if you are standing still.

When you get home, repeat this procedure in the garden, taking the puppy into the garden first. Then

Cats and puppies need careful introduction with puppies being restrained and cats being allowed to climb up high if they want to.

let your puppy into the house, bringing the older dog in afterwards. Pick up any toys, bones, food dishes, beds and blankets beforehand. Keep the excitement to a minimum and try not to interfere too much. If you are worried that your older dog may attack your puppy, put the puppy in a playpen or use a stair gate so that both can investigate one another safely.

When they have settled down, replace the beds and blankets, but keep any toys and bones out of the way for a few days. Try to give the older dog more attention and affection than you did before the puppy came, and remind the children to do this, too.

Introducing cats

Usually introductions between very young puppies and cats go quite smoothly. When introducing your puppy to an adult cat, restrain the puppy, not the cat. If you have a cat that stands up for itself, it will probably hiss and spit at the puppy to begin with, and the puppy will retreat. If your cat is the sort that will run away from the puppy, be ready to hold and distract the puppy so that he is not tempted to give chase.

Throughout your puppy's first year, ensure he cannot give chase at any time. Continue to restrain him until the cat is accustomed to his presence and the puppy has learned that the cat is not for chasing. During times when you are not there to supervise, keep your puppy in his playpen (see page 45) or use stair gates, so that the cat can stay away from this area if it wishes.

Problems may arise later when the puppy is more confident. Often he will begin to bounce at the cat, having learned that this makes the cat run. If you can see this beginning to happen, interrupt your puppy and offer a game with a toy instead. By doing this, you will be teaching him that cats are not for playing with and, instead, that humans are the source of all games. If necessary, attach a line to the puppy's collar and quickly stand on it to prevent your puppy chasing the cat whenever you see him preparing to do so, and then distract him with a toy. Giving the cat plenty of places up high in the house so that it can get to a safe spot without running will also help.

The first night

It is helpful to make your puppy as sleepy as possible before he goes to bed. If he has had a long journey and has spent all day with his new family, the chances are that he will be tired anyway. Playing with him will help to use up the last of his energy, and feeding a warm meal will also help. Remember to take him out to the toilet for one last time before he finally settles down.

Before you collect your puppy from the breeder, take a small blanket and ask them to put it into the bed where the mother and puppies sleep for a few days (it gets very dirty and smelly!). Collect this with your puppy and seal it in a plastic bag. When your puppy goes to bed for the night, wrap this blanket around a warm hot-water bottle and put it into your puppy's bed; he will find this very reassuring and comforting because it smells familiar.

Opinions vary on how you should handle your puppy's first night with you. Puppies howl or cry when they are separated from their mother and litter-mates. Being isolated is not a natural state for

Keeping your puppy close at night at first allows easier housetraining and prevents stressful, sleepless nights.

animals that live in packs, and my view is that they should learn this gradually. During the first few nights take him up to the bedroom with you, and put him on a blanket in a high-sided cardboard box that he cannot climb out of. Any whimpering can be quietened by a reassuring pat or a quiet word. Do not over-fuss or respond to every murmur.

If he wakes up, cries loudly and tries to get out in the middle of the night, he probably needs to go to the toilet. Get up (even if it is 3 a.m. and raining) and take him quickly outside to the garden. Praise him if he goes to the toilet, and take him back to his box until you are ready to get up (see page 70 for housetraining procedures).

After the first few nights, your puppy should be settled into your home and should have become accustomed to being without his mother and litter-mates. He will also have got used to being left alone during the day (see page 154). Consequently, by the end of the first week in your household he should

Confining your puppy in a playpen when you are busy and cannot supervise prevents bad habits forming.

Make sure your puppy has plenty to do while he is in the pen and don't leave him there for too long.

be ready to sleep on his own at night. He may be a little unsettled for the first few nights in his new bedroom, but do not go in to him if he howls or cries when you shut the door. Before leaving him for the night, it is useful to put down a large sheet of polythene covered with several sheets of newspaper, in case of accidents.

Some puppies will not want to go to the toilet in the house at night and, on waking up, will make a lot of fuss to be let out. If he does this, it is worth going down to let him out, rather than force him to go on the floor. Do not give him any attention while you are

down there; you are only there to open the door and accompany him outside. Otherwise he will begin to cry for you whenever he wakes up and finds himself alone. Only when your puppy relieves himself outside should you give him attention and praise for doing the right thing.

A puppy playpen

If you have a busy life, this is an essential piece of equipment for bringing up a well-behaved puppy. Not only does it allow you to relax mentally and forget about your puppy for short periods – essential if you are not to get over-tired and cross with him – but it also teaches him some self-control. Dogs in a human household have to learn to lie down and relax when no one wants to play or give them attention. In the playpen there are limited options for activity and they quickly learn to settle down.

Equally important is a playpen's role in preventing puppies from getting into all sorts of trouble while their owners are busy with something else. Puppies can learn many bad habits if they roam the house unchecked – such as chewing electricity cables or shoes, and stealing biscuits from the coffee table – especially if left at large when there is no one at home. Since you are not there to supervise, bad behaviour goes uncorrected, and behaviour that you want is not encouraged.

A simple way to prevent this is to construct a puppy playpen. This can be as large as you like, but should be at least large enough to have a sleeping area and another part, covered with newspaper, where your puppy can get out of bed to play, move around and go to the toilet if necessary.

You can achieve this by barricading off a corner of the room in some way. Make sure the partitions are secure and cannot fall on your puppy or trap him if he tries to escape. Alternatively, you can buy specially made pens, often made up of panels that are linked together. You can order these from good pet shops. For some puppies, you will need a lid to prevent them from climbing out when they get older.

The best place for the playpen is usually the kitchen. In most households this is the place where people congregate and pass through, and where there is usually something happening. Your puppy can then get used to many different sights, sounds and smells from the safety of his playpen. In addition, kitchens generally have a washable floor, which is useful for housetraining accidents. If you have a small kitchen, it is better to give the puppy a smaller pen than to put him in a larger one in a room where few people go.

Put your puppy's bed in the playpen with one or two chews and toys. He can then be safely left in there whenever you go out, or whenever you cannot concentrate on him. When your puppy is allowed out, you are there to teach him right from wrong. The training process is then much quicker because your puppy is never rewarded for unacceptable behaviour and will rapidly learn to show only the behaviour you want. This is kinder to the puppy and easier for you. Once your puppy has learned right from wrong and behaves well in the house (even when left alone), you can dispense with the pen.

Not a punishment

Do not use the pen as a prison when your puppy has done something wrong. If he does something you do not like, simply correct him, show him what you want him to do and praise him for being good. The pen should not be associated with punishment. Talk frequently to your puppy when he is in the pen and, if it is big enough, play with him in it, so that he enjoys being there.

Do not keep your puppy in the pen for long periods. It is meant only as a safe place to keep your puppy while you are engaged elsewhere. He should be given as much time and attention outside the pen as you can manage, and should not be left there for more than two hours and certainly not all day.

Introducing the pen

If you introduce the pen at a very early age, your puppy will accept it as part of life. Encourage him to go into the pen by running towards it with him and throwing treats into the pen through the doorway. Calmly close it behind him while he finds and eats the treats.

If you have an older puppy, start by leaving the pen open, putting his bed inside and encouraging him to go and rest there when he is tired. Throw food treats and toys into the pen, and praise and play with him whenever he goes in there of his own accord. It is much better if the puppy goes in of his own free will rather than being forced to do so each time. After a few days, when he is happy to go inside, you can begin to confine him there whenever you need to.

Barking in the pen

Barking in the playpen is a sign that you have left him there for too long. Do not, on any account, take your puppy out of the pen, tell him off, talk to him or look at him if he barks or whines. Do not pay him any attention at all until he is quiet again. If you do, you

Encourage your puppy to go into the pen by throwing a few tasty food treats inside for him to find.

will be reinforcing this behaviour and he will quickly learn that he can get his freedom by barking.

After a few minutes of quiet, let him out to go to the toilet or for a game – but it should be your decision, not his. The only possible exception to this is if you think he may want to go to the toilet and is asking to go outside. If this is the case, wait until the barking or whining ceases and then whisk him outside quickly. Otherwise, make it a rule that he only comes out of the pen when he has been well behaved.

A retreat from children

If you have young children, your puppy can easily become over-excited and play sessions can get out of hand. Using a playpen means that when the puppy is allowed out, you are there to supervise, teach 'good' games and prevent play-biting. A playpen also ensures that the children do not tease the puppy or teach him bad habits, and it can also be a useful means of teaching him not to chase running children.

An indoor kennel can be a safe haven for your puppy if covered to make it den-like, but leave the door open at all times.

Puppies need quite a lot of sleep, especially when young, and an over-stimulated, over-excited and over-tired puppy is very likely to become irritable and snappy. Judicious use of the pen can prevent this. Your puppy can be put there to rest periodically and, once he has learned that he cannot get out to play, will quickly settle down and sleep.

Your puppy also needs to learn that he cannot join in all of the children's games and that, sometimes, he has to sit quietly while they play.

The playpen is an ideal way of achieving this. Some puppies will want to go to the pen to rest when they are tired and the children become too much. Leave the pen open, if possible, and make sure the children know it is a no-go area.

Indoor kennel

If you do not have enough space for a playpen, or cannot construct one, an alternative is to use an indoor kennel or travelling cage. This is a small mesh cage (it should be as large as possible, and at least large enough to enable your puppy to stand up, lie down and turn round comfortably once fully grown).

The big disadvantage is that there is nowhere for your puppy to go to the toilet, and he cannot exercise or explore in it. This means that you cannot leave your puppy in it for longer than an hour at a time, so its use is limited. However, an indoor kennel is better than having nowhere at all to confine your puppy while you are involved in something else. Just be careful about the amount of time your puppy is left in

it; make sure he has chews and toys to play with, and do not put him there unless he is tired and sleepy and has recently been to the toilet.

Using a stair gate

Stair gates (or baby gates) can be useful, especially if you have young children as well. They can be used when you do not want your puppy to be in the same room as you – for example, when you are playing with your baby on the floor or have important visitors that you want to concentrate on. The puppy can still

A stair gate prevents your puppy following you out the room, but still allows him to see what is going on.

have your company, but can be safely kept away from whatever is going on, by putting him behind the gate in the doorway.

Setting a routine

It is important to establish a routine for your puppy. He will adjust to life in your household much more easily if there is some sort of order to his life in the early weeks. Keeping to a routine will also be easier for you because it allows you to cope with everyday chores, such as housetraining, without having to think too much about them. Having a written routine to follow may help to ensure that, in a busy household, your puppy's needs are not forgotten.

Feeding your puppy at regular times will make the housetraining process easier.

SUGGESTED DAILY ROUTINE	
MORNING	
8 a.m.	Wake up; out to toilet; short play session
8.30–9 a.m.	Family's and puppy's breakfast
9.15 a.m.	Out to toilet; short play session; rest period
10 a.m.	Out to toilet; play/training session; rest period
11 a.m.	Out to toilet; socialization time
AFTERNOON	
12 p.m.	Out to toilet; family lunch
1 p.m.	Puppy's second meal
1.15 p.m.	Out to toilet; rest period
2 p.m.	Out to toilet; short play session; rest period
3 p.m.	Out to toilet; supervised freedom of the house/play time
4 p.m.	Out to toilet; rest period
5 p.m.	Puppy's third meal
5.15 p.m.	Out to toilet; rest period
EVENING	
6 p.m.	Play/training session; out to toilet
7 p.m.	Family dinner; rest period
8 p.m.	Out to toilet; socialization time; handling/grooming session
9 p.m.	Puppy's fourth meal
9.15 p.m.	Out to toilet; play/training session/supervised freedom of the house
10 p.m.	Out to toilet; vigorous play session
11 p.m.	Out to toilet; bed

Regular outings to the garden are essential if the puppy is to learn to be clean indoors.

The routine on the opposite page is for a very young puppy. You may find that this plan does not suit your household or your puppy, but it should give you some ideas to enable you to design your own. The routine is a basis for what happens each day. It is flexible, to enable you to live your own lives around it, but it should be kept to as much as possible to allow for housetraining and feeding and to enable you to make time each day for essentials, such as play, socialization and grooming. The routine may look complicated and time-consuming, but it will make everything easier in the long run. For example, it may seem unnecessary to take your puppy out to the toilet each hour. However, taking him out regularly is easier than cleaning up accidents, and he will learn in a much shorter time.

Introduction to collar and lead

When your puppy has had a few days to settle into your household, it will be time to get him used to

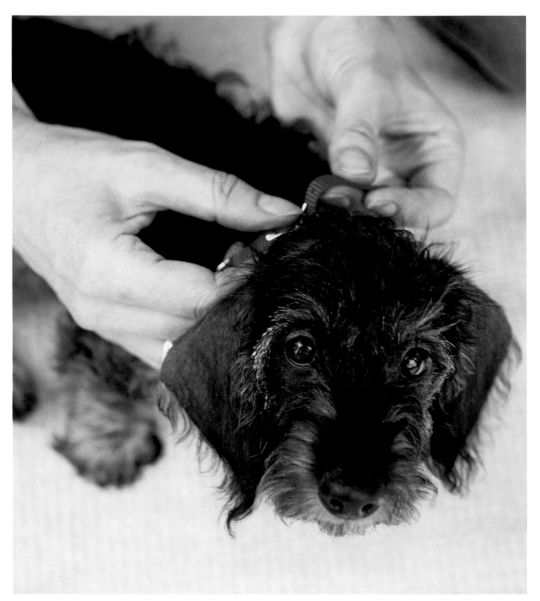

A collar will feel strange to the puppy at first but he will soon get used to it.

wearing a collar. Buy a suitably sized buckle-collar (never a check-chain or half-check collar). It needs to be large enough to enable you to insert two fingers between the collar and his neck when put on.

Place the collar on your puppy just before something pleasant is about to happen to him – for example, you are about to play with him, take him outside or feed him. He will probably attempt

frantically to scratch it off at first. Ignore this and praise him as soon as he stops. After a few moments, distract his attention from his collar with the next event. Take the collar off during this event and put it on again later.

After a few days of this, he will get used to the collar and begin to ignore it. It can then be left on all the time. Puppies can easily get lost, so attach an identification tag to it. Remember that puppies grow at a tremendous rate, so check the collar every few days to ensure it is still fitting well.

Later, as your puppy becomes used to being restrained and handled, begin to accustom him to being gently restrained by his collar. Initially hold his body with your free hand to prevent him from swivelling round and trying to pull away from you, twisting your fingers in the collar as he does so. Ignore any wriggling, and praise him when he stands still. Gentle, firm restraint in this way will soon teach him that he cannot get away if someone takes hold of his collar, and he will learn to accept it.

When he is used to being restrained by the collar, attach a lead and let it drag around after him during a few play sessions so that he becomes used to the feel of it. Pick up the end of the lead sometimes, but keep still when you do this. Your puppy needs to learn that being on the lead means that he is fastened to his owner and cannot go where he wants to any more. When your puppy has accepted the restraint, praise him and let him go free again.

Teaching your puppy to accept being held and restrained by the collar is an important precursor to attaching the lead.

CHAPTER 6
Socialization

One of the most important things that can happen to a puppy during his early life is socialization with all the different types of living creature he will meet in future, as well as habituation to all the noises, smells, sights and experiences in his new world. Puppies that grow up without such good experiences go through life afraid, wary and more likely to be aggressive. Puppies that have many good experiences with everything that will surround them in future grow up confident and happy. The advice given in this chapter is probably the most important offered in this book if you want your puppy to grow up well adjusted and friendly. To make sentences in this chapter less convoluted, the term 'socialization' is used both for socialization (getting used to the living part of the world) and for habituation (getting used to the non-living part of the world).

Key point
- If you do nothing else, you owe it to your puppy to find time for socialization. It is not difficult to do, but it needs to be done at once, while your puppy is still young – not tomorrow or the next day, but today. Socialization is probably the most important factor in the future well-being of your dog and in the formation of a well-balanced, friendly adult.

Lifespan
Puppies take a relatively short time to grow up compared with humans. Each precious week during puppyhood is equivalent to approximately five months of human development, so much time and effort are needed during a puppy's first year to ensure he is raised correctly.

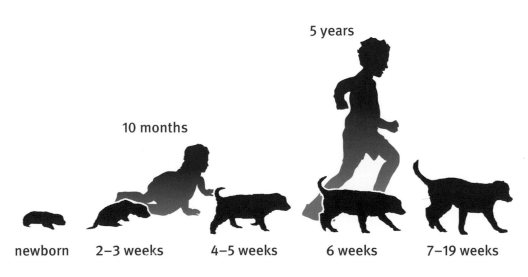

5 years

10 months

newborn   2–3 weeks   4–5 weeks   6 weeks   7–19 weeks

Life in the wild

In the wild, pups can learn about their surroundings more quickly if they have no fear, and it makes sense for them to be unafraid while they are still under the protection of their mother and other members of the pack so that they can explore easily. As the pups mature and wander off to explore on their own, it is important for them to be cautious of situations and objects they have not been familiarized with when young. Domestic dogs have inherited this delayed-onset fear of new experiences to a greater or lesser extent, depending on which breed they are; and, as they get older, puppies become more cautious. Anything that has not been encountered before will be met with suspicion and fear.

When to socialize

One of the major differences between humans and dogs is the different speed at which we grow up. It takes a human approximately 18–21 years to mature (although I have known it take longer!), whereas dogs mature in about one to one-and-a-half years. A week in our childhood is a relatively short period,

whereas to a puppy it represents a large proportion of his puppyhood.

While a puppy is very young, new experiences are approached without fear. As they grow, puppies become increasingly fearful of new encounters and, by just 16 weeks of age, the window of opportunity closes as fear starts to outweigh the desire to approach. It is important, therefore, that you make the most of early puppyhood and pack in as many good experiences as possible during this time. Deep and lasting impressions are formed at this age, and whether these experiences are good or bad will be remembered throughout a dog's life.

Socialization should be a continuous process up to the age of maturity, and concentrated effort should be put in until this age. The more new experiences and enjoyable encounters that a puppy has as he grows up, the more likely he is to mature into a well-adjusted adult capable of taking anything in its stride without becoming fearful. Socialization becomes progressively more difficult as the puppy grows older, but it is still worth doing. Puppies that have missed out on early socialization will need extra effort to help

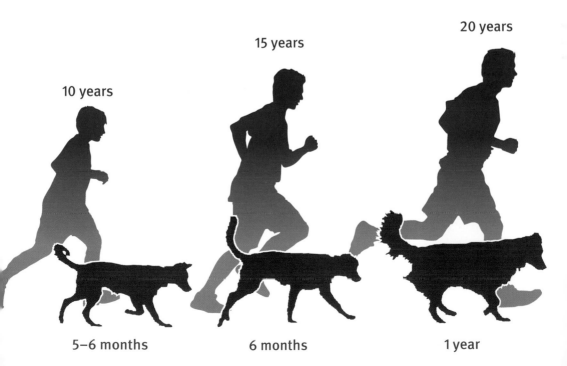

10 years

15 years

20 years

5–6 months

6 months

1 year

them catch up, but you will need to go slowly and be patient while they overcome their fears.

The breeder's role

By the time your puppy is weaned and ready to leave his mother, it is already nearly halfway towards the closing of the window of opportunity, and the process of socialization should have already begun in earnest. When the puppy is ready to go to his new home at six to eight weeks, about one month of prime socialization time has gone by. The breeder therefore has a responsibility to ensure that puppies are well handled and socialized during that time. This is why you need to make a careful choice about where your puppy comes from (see page 16). If you obtain a puppy from a breeder who has not done this adequately, you need to make up for lost time and work very hard at your socialization programme.

Vaccination versus socialization

The usual ages for vaccinations to be given are eight and 12 weeks (sometimes 16 weeks as well), but puppies can now be vaccinated at six weeks of age with a reasonable chance of success. Until the vaccinations have taken effect, your puppy will be at risk of contracting diseases from other dogs.

Early vaccination is strongly recommended since the benefits of early socialization are so great. Talk to your veterinary surgeon about the options available. Their knowledge of the latest vaccines and local disease conditions will allow you to make an informed decision. My view is that a compromise is necessary between protection of physical and mental health. Early vaccination at six weeks will give you the best chance of protecting your puppy before the later vaccinations take effect.

Taking your puppy out and about, but taking reasonable precautions to ensure he is not exposed to disease, will allow him to socialize before he is fully covered. Unless he is very large and heavy, carrying him outside to experience the world is a better option than keeping him inside. Put him down in places where no other dogs have toileted, keep him away from unvaccinated dogs, and give him the opportunity to see as much of his new world as possible at a young age. Otherwise you will lose weeks of vital socialization time, which could make the difference between a shy, fearful adult and a well-adjusted, friendly one.

How to socialize successfully

Try to make each new encounter a happy, positive experience. Make sure your puppy's tail is relaxed and wagging. If he is moving forward, enjoying the experience and having fun, he will be storing away those good encounters for the future and learning to feel safe and comfortable in his world.

The secret to good socialization is to try to look at the world from your puppy's point of view. Try to imagine what it is like to be that small and vulnerable and to have such a limited knowledge. Do not overwhelm him with too much all at once, but start with brief encounters and slowly build up until he is experiencing more and more. Ensure that he enjoys each new experience, playing with him and talking happily to him, so that he knows there is nothing to be worried about. His natural curiosity will make him want to explore new situations and, provided you protect him from becoming afraid, he should become more and more confident as he gains experience.

Meeting and having pleasant encounters with children of all ages is essential socialization for your young puppy.

Carrying your puppy into the outside world will allow him to experience new situations from a safe vantage point.

Try to think ahead to protect him from unpleasant or scary experiences. You will need to think about the way he is viewing the situation and, if necessary, modify his experience so that he does not become fearful. Carefully observing your puppy's reactions will help you to tell if he is even mildly fearful. Watch his ears and tail and his body posture. Is he tense or trying to make himself smaller by lowering his tail and ears and by crouching? Some puppies will not show these signs so obviously, but will try to escape from the source of the fear if at all possible. Is he straining to get away, but being restrained by his lead? By carefully watching your puppy, you will be able to tell how he is feeling and make changes if you think he is overwhelmed or scared by the situation.

It is important to allow your puppy to approach in his own time. Do not use the lead to try to drag him towards people or into new situations. Remove the pressure by talking to the people for a while and let him make his advances when he wants to. Games with toys or feeding treats can help to speed up the process and enable your puppy to feel at home with the new situation more quickly. Give visitors treats to hand to him, or get them to offer a game with a toy, and they will often be accepted more readily. If your puppy will not play or eat in a new situation, he is probably feeling too much anxiety and you need to do something to help reduce it. If he is just at the edge of what he can cope with, playing with him and feeding may help him to relax.

Socialization **57**

Making friends with strangers is important if your puppy is to be well adjusted. Use treats or food to encourage the puppy to approach.

If your puppy is showing signs of apprehension or is already fearful, tone down the new experience so that he is no longer afraid. In practice, this can mean putting more distance between yourself and the situation, or removing one of the new elements. For example, if your puppy has not met children before, four noisy children may be overwhelming so restrict the encounter to one quiet child at first.

Try not to act sympathetically if your puppy becomes apprehensive. Showing your concern will cause him to think there is something to be afraid of. Instead, try to jolly him along and look as if you are having a good time and are not worried. At the same time, change the situation so that your puppy has less to be apprehensive about, by moving away or distracting him into thinking of something else.

When you know your puppy's limits, extend them each day. Expose him to as much of a new situation as he can cope with without becoming overwhelmed or tired. Encourage him to enjoy the experience with food and games and he will gradually become more confident. The more situations your puppy is familiar with and the more happy experiences he has, the less time he will take to adjust to new encounters.

There is a socialization programme for your puppy in the appendix (see page 200). Use it, or design your own, so that you have a structured plan to work with and keep to. This will help prevent you from putting off socialization sessions in favour of something you want to do instead. Your puppy should come first at this time – you can never get this time in your puppy's life back again.

Take your puppy out every day to different places so that he has many encounters with a wide variety of situations and environments.

Socializing with humans

Humans come in all shapes, sizes, ages and types of character. Your puppy will rapidly get used to people he sees every day or who regularly visit your house, but he will be unfamiliar with others and may be afraid of them in later life unless you take him out to meet them.

If you want your dog to be 100 per cent confident of all humans, you need to introduce him to all sorts of people and he needs to make at least ten good friends with different people. Inviting friends to your house while your puppy is still young is one of the best ways to socialize a puppy with humans. If he is shy, allow him to come out in his own time and do not try to force the issue or you will make him worse. Prevent well-meaning people from approaching him too quickly; your puppy should make the first move. Smelly, tasty food treats given by the visitor can help to overcome any fear he may have.

Humans are also prone to touching unfamiliar dogs, especially if they own dogs themselves. Handling exercises to do with your own puppy are covered later (see page 136), but it is important to ensure that your dog is happy to allow strangers, both adults and children, to touch him all over. Teach him to trust you first, then ask others to handle him too.

Teach your puppy to greet people nicely at the front door by using a lead to prevent any jumping or too much exuberance.

Pleasant experiences with small children will help your puppy to feel comfortable with them and make him much less likely to bark or snap during any worrying encounters later on.

Later, invite people who are not keen on dogs or who do not know how to handle them as well as those who do. People who are unfamiliar with dogs tend to approach them in very different ways from those that are familiar with them, and it is wise to accustom your puppy to all types of people.

Exposure to people needs to be both at home, on your own territory, and off territory in parks, towns, on the street and in other people's homes. Your puppy needs to get used to people wearing glasses, beards, unusual clothing, uniforms, hats, motorcycle helmets, carrying bags or sacks, people in wheelchairs, people who walk with a limp or carry a walking stick, and people of different races. He also needs to get accustomed to people running or jogging, playing football, riding bikes and all the other strange activities that humans engage in!

Until your puppy has learned to enjoy the company of people, ask them to look away if they are frightening him by staring at him (even if they are doing it unintentionally), or distract their attention from him by speaking to them. This will take the pressure off the puppy and give him a chance to relax. In time, your puppy's natural curiosity will encourage him to come forward and make friends.

Getting used to children

It is particularly important to get your puppy used to children of all ages, especially if you do not have children of your own. Even if you do, he will need to accept other people's children who are different ages from yours and have different characters. Try to ensure that your puppy meets all ages of children – babies, toddlers, infants, juniors and teenagers.

One of the easiest ways to do this is to walk your puppy to the local school and back at the appropriate times, even if you do not have children yourself. Children are naturally attracted to puppies, which makes it easy for you to get the two of them together, but do take steps to ensure that your puppy is not overwhelmed by the experience, should the numbers of children get too great. Again, giving children tasty treats to offer to him can make the experience very pleasant for your puppy, and he will begin to look forward to meeting them (show them how to offer treats on the flat of their hand, so they do not get their fingers bitten). Allow him time to get to know one or two of them so that he accumulates friends of different ages. Take toys with you, so that they can play with him as well.

Taking your puppy to children's play areas allows him to become used to the sight and sound of children playing, running and screaming. Teaching him to remain calm in these situations and not to join in is a valuable lesson. It will help to prevent later misunderstandings when your dog may run after and jump up at a child in play, possibly bringing the child to the ground. Being chased by a dog, even if the dog is just playing, can be a very frightening experience for a child, particularly if they are not used to dogs or if the animal is larger than they are.

Make sure your puppy has been exercised and allowed time to go to the toilet before you get to the children's play area, and clear up after him if he should mess in the vicinity. Talk to your veterinary surgeon about an adequate worming programme for your puppy so that there is no public health risk.

Socializing with other dogs

Great care should be taken when introducing your puppy to adult dogs. Do not do so unless you know the dogs have been well socialized themselves and are used to meeting unfamiliar puppies.

This does not, however, mean that you should keep your puppy away from all dogs. Doing this could be just as harmful, because he will then be unfamiliar with other dogs and uncertain of how to act around them. This uncertainty could lead to aggression later on. Instead, try to find a number of other dogs that you know will be good with puppies and introduce them on neutral ground, in an area where both can be safe and free to investigate each other and their surroundings.

Take care when walking your puppy in places where other dogs are loose. A confident dog running up to a shy puppy can cause a big fright, even though it means no harm. If the puppy is restricted by a lead he may be more frightened, or if he is loose he may run away, which could be worse. A fright like this can leave a permanent mental scar. Protect your puppy from such encounters by keeping your distance, and encourage him to play games with you instead.

Try to make sure that all the experiences your puppy has with other dogs are positive, then he will have no reason to be afraid of (and possibly aggressive towards) them later on.

Puppy socialization and training classes

One of the best ways for puppies to build on their knowledge of dog communication systems, which they began in the litter, is for them to attend well-run puppy socialization parties or training classes (see page 193 for advice on finding a good class). Socialization parties are often run for puppies of 10–12 weeks of age at veterinary surgeries, and puppies move on from these to puppy training classes between the ages of 12 and 18 weeks.

Look for parties and classes where puppies are allowed to play with no more than two other puppies at a time, carefully selected to match their own age, size and experience. It is important to avoid parties or

Learning to be respectful to adult dogs is important for your puppy, so try to find well-socialized dogs that are good with puppies for this purpose.

Puppy socialization classes will enable your puppy to play with others his own age.

classes where all puppies play together all the time. This can quickly lead to shy puppies becoming aggressive and to confident puppies learning to bully.

All training at the class should be done using reward methods only. Before joining, it is sensible to go to watch a class in progress without your puppy, to see if you approve of the methods used. By attending puppy classes, your puppy will continue to learn the body language and signals necessary for adequate communication with other members of the species in a low-risk environment. You will learn how to train him and how to understand him better.

Contact with adult dogs is needed in addition to these classes. Puppies should learn that not every dog is available for play all the time. A friendly adult dog that likes puppies but will not accept rough play will firmly teach your puppy respect for others without becoming aggressive, although finding suitable dogs can be difficult.

Socializing with other animals

If young puppies meet a variety of animals before they are 16 weeks old, they will accept them later as part of their social network. Getting puppies used to other animals later on is a little more difficult, but well worth doing before they reach maturity. Try to accustom your puppy to anything he will be exposed to in later life, but keep him under control so that he cannot give chase or show unwanted behaviour.

Getting your puppy accustomed to cats and horses is useful – as well as to livestock if you live in or near the country. Try to make sure that he sees so many of them while he is under control that he becomes bored with them. If a puppy frequently sees livestock and is prevented from giving chase, he will learn to accept that they are boring creatures that do not run. If he walks past many cats in their gardens while he is on a lead and is not allowed to frighten them, he will learn that they do not run and are not for chasing.

Similarly, if horses routinely walk past him while he is a puppy and he is prevented from giving chase, he will get used to them.

New situations, objects and events

Puppies need to get used to all the noisy, smelly, strange-looking things in their new world. This includes everything from household machines, such as vacuum cleaners, washing machines and hair dryers to things they might encounter outside, such as traffic, joggers, skateboards and pushchairs.

Things that seem so ordinary to us, such as cars, bicycles and lorries, can seem like huge roaring monsters to a new puppy out for the first time, and any fear or apprehension he may have will need to be overcome. Give your puppy time to approach things when he feels comfortable, making it fun to be at a

distance until he is ready to move closer. Take your puppy into all types of situation that may be useful later on. If you do not have a car, try to borrow one so that he becomes used to journeys in it. If you do not normally use public transport, make a point of taking him on buses and trains, which could be very useful later on when the car breaks down! Try to arrange for your puppy to experience new sights, sounds and smells every day.

It is important to get him used to a variety of sudden-onset noises, such as thunder and fireworks, early in life so that he does not develop noise phobias later. An easy way to do this is to purchase one of the many CDs specially designed for this purpose. Play it quietly at first and combine it with happy events for your puppy, so that he has good associations with these noises. Work up gradually to

Your puppy will rely on you to guide him through new experiences and show him how to behave.

playing it louder, until he is happy to tolerate loud bangs and thundering because he is waiting for something nice to happen to him.

Try to go everywhere and do everything that you may want to include your dog in when he is older. Visit a dog-friendly restaurant, bar or café, and take any opportunity to go to small dog shows and fairs where your puppy can have a wide variety of different experiences. If your puppy shows apprehension of anything – whether it is objects, situations or events – go more slowly and work patiently with him until any fear disappears.

What to do in a bad experience

If, despite your careful supervision, your puppy does have a bad experience, it is important to put this right as soon as possible by arranging for your puppy to encounter the same situation again, but this time for it to be a pleasant event. Several of these pleasant encounters will be needed to overcome the unpleasant one. How many will depend on how shy your puppy is; timid puppies will need more pleasant events to overcome an unpleasant experience.

For example, say a large Labrador comes running over to see your puppy. He is being friendly, but your

Vacuum cleaners are noisy, smelly and can be scary. Let your puppy get used to them slowly by keeping brush movements slow and deliberate.

Steps can be daunting for small puppies, but they soon learn how to negotiate them and take them in their stride. Going up is often easier than coming down!

This shy puppy sniffs the ground while keeping an eye on his surroundings. He'll need many pleasant encounters if he is to grow up well adjusted and unafraid.

puppy is very scared and tries to run away, but is prevented from doing so by the lead. This type of situation can result in your puppy being scared of other dogs that approach quickly, possibly leading to aggression in later life, and it needs to countered with many happy experiences as soon as possible. Try to set up a situation with another similar Labrador, this time on a lead and under control so that it approaches more slowly, giving your puppy time to relax and make friends. If your puppy is still worried, allow him to approach the Labrador in his own time until he builds up his confidence. Continue until your puppy is happy to greet any dogs that approach quickly.

Taking the time to make up for bad experiences will result in a puppy that is comfortable with the world

he lives in, and there will be no unpleasant memories stored that may trigger an unexpected aggressive event in later life.

The shy puppy

How shy a young puppy is, and hence how much socialization he requires, will depend partly on his genetic make-up and partly on the environment in which he has found himself during his short life.

Fearfulness is an inherited trait, and your puppy will have inherited the predisposition to be afraid of the unfamiliar to a greater or lesser extent, depending on the breed to which he belongs. Reactive, sensitive breeds, such as those bred to herd, are more shy and need much more early socialization than more mentally robust breeds, such as gundogs.

If you have a quiet family with few comings and goings, your puppy is likely to be more shy than one that lives in a very lively household. Whether or not you have taken your puppy out in the early weeks since you had him, or whether you kept him at home, will also have an effect.

Whatever his background, you need to be aware of how shy your puppy is and watch him while he encounters situations for the first time (see page 26 for information on how to read your puppy). Is he moving from or towards an object or person? Is he looking around calmly, or does he orientate to any new sound rapidly and with alarm? If you own a shy puppy, it would be wise to invest a lot of time in socialization as soon as possible. A dog is never too old to socialize, but it does become progressively more difficult as he grows older.

If you find something your puppy is afraid of, try to overcome his fear gradually. For example, if he is afraid of passing cars, find a place where you can stand some distance from the road. When you sense a little apprehension, stop and play a game with your puppy (keeping him on his lead), feed him food treats, talk to him and excite him until he is happily playing and having a good time. Repeat as often as possible over a few days, getting closer to the road each time until he is not afraid to stand at the kerb. Never go beyond your puppy's limits.

Similarly, if he is afraid of people, arrange for him to make friends with gentle, quiet people before moving on to more boisterous ones. If he makes friends with about ten different people, you will find that he starts to generalize and expect the next person to be friendly, too. Given time and the right experiences, even very shy puppies can begin to relax and enjoy the company of strangers. You will need to work extensively on a very shy puppy at a speed set by him. Make sure that he is happy with all types of people approaching him, calling his name and staring at him before you begin to relax his socialization programme.

A sideways orientation of the body and an offer of a game with a toy draws this
shy puppy closer.

CHAPTER 7
Housetraining

Most animals that are born in a nest have an instinctive desire to move away from the nest to go to the toilet. They will do so, without being taught, as soon as they are able. Dogs are no exception, and at the age of about three weeks they will begin to leave the sleeping area to urinate. They are, as it were, pre-programmed to be housetrained; we just have to teach them that houses are our nests and that they have to go right outside when they want to go.

It is important for one or two people in the household to take responsibility for housetraining.

Puppies need constant supervision for the first two to three weeks until they learn good habits. Being vigilant, keeping your puppy in view at all times and concentrating on making sure he goes outside at the right time will give you a clean puppy that has very few accidents.

Take your puppy outside to the same spot in your garden at the following times:

• Shortly after each feed
• After playing
• After exercise
• After any excitement (e.g. visitors arriving)
• Immediately upon waking
• First thing in the morning
• Last thing at night
• At least once every hour.

Allowing him to wander around and sniff at the ground will help to speed up the process. Do not pressure him to go, but it is really important to stay out with him. Take a coat if it is cold or an umbrella if it is raining so that you are not in a hurry to go back in.

Be patient and wait. When you see signs that he is about to begin, say a chosen phrase to him. Choose anything you can say easily in a public place (you may need it one day when people are listening), such as 'Be clean!' When he has completely finished, praise him enthusiastically (make that tail wag!) and play a game with him. Keep the area clean by picking up any mess and burying it or flushing it down the toilet.

Puppies are easily distracted when outside so having the patience to stay with him until he has settled down is essential. If you leave him to it, he will probably run to the back door and spend the rest of the time trying to get back in with you. Once you let him in, the stress of the separation, together with

Encouraging your puppy outside regularly will speed up the housetraining process.

Be prepared to go out in all weathers and stay out with your puppy while he goes. Having a warm coat and umbrella ready will help, especially for visits to the garden in the middle of the night.

the increased excitement and exercise, will make him want to go and you will be left with a mess inside and an uneducated puppy. There is no need to stay outside for hours, waiting for him to go. Wait for a few minutes only and, if nothing happens, take him inside and try again a little later.

If, at any time of the day, you notice your puppy walking uncomfortably with a far-away look in his eye, sniffing the floor and circling or getting ready to squat, immediately interrupt him and take him outside. Let him walk. Do not pick him up or he will not learn the vital link in the process, which is: 'When

I need to go, I need to get to the back door and into the garden.'

If, at any point, you catch him in the act of going in the house, shout to interrupt him. What you shout is immaterial, but it needs to be loud enough to capture his attention and stop him mid-flow, but not so loud that he runs for cover. As soon as you have shouted, run away from him, towards the back door, calling him happily and enthusiastically to encourage him to follow. Go outside to your chosen spot and wait until he has relaxed and finished what he started earlier. Say your chosen phrase, praise him and play with him when he has finished, as usual. Take him back into the house and put him in another room while you clean up any mess.

At times during the day when you cannot concentrate on your puppy, it is best to keep him confined to a smaller area where accidents are not too important. If you are using a puppy playpen, cover the floor with one large sheet of polythene with newspaper on top so that accidents can be cleared up easily. If accidents do occur, consider it your fault and take your puppy out more frequently.

By following these simple procedures, your puppy will quickly learn that the place to go to the toilet is outside and will get into good habits. The more frequently you take him out at the appropriate time and the fewer times he goes indoors, the quicker he will learn. Regular habits take time to develop, though, so be prepared for the occasional accident.

Once your puppy has formed the correct habit, he will begin to show signs of wanting to go outside whenever he feels the need to go. Watch for the telltale signs, such as running to the door and back, whining at the door or pawing at it, and let him out immediately.

Accidents will happen

Toileting inside will leave traces of scent on the floor that your puppy, with his ultra-sensitive nose, will be able to detect long after you have disinfected and cleaned the area to your satisfaction. This will encourage him to use the spot again.

To prevent this, use one of the following to clean the area:

- Special products available from veterinary surgeons for this purpose
- Hot biological washing-powder solution; allow it to dry, then wipe over it with alcohol (such as surgical spirit); do a patch-test first to ensure that any colours in the carpet do not run.

Keep an eye on your puppy and watch for signs that he may need to go, such as sniffing, moving in circles and looking 'distracted'.

Stay outside with your puppy. Otherwise, he will turn his attention to getting back in with you and will still need to go when he's back inside.

What to do at night

Puppies, like children, have only limited control over their bodies and, in general, when they feel the need to go to the toilet, they have to go straight away. Expecting them to last through a six- to eight-hour night-time is too much, and puppies may be seven or eight months old before they are completely clean. If you have a puppy that wakes you up when he feels the need to go out rather than mess on the floor, it is well worth getting up. He will be housetrained more quickly because he is not making mistakes and it will not be too long before he can hold on all night.

Once you begin leaving him on his own all night, cover the whole area with a sheet of polythene and newspapers, to make sure no messes are made on the floor. The polythene will stop any waste matter leaking through, thereby preventing smells on the floor which would encourage this habit during the day.

Not clean at night?

Many young puppies ask to go out during the day when their owners are there, but are not clean during the night, or if left for any length of time during the day. This is because immature animals need time to develop full control over their bodies and, when they need to go, simply cannot hold on until you come back. Owners often worry about this and think their dog will never be clean when it is left.

There is no need to worry. Puppies take different lengths of time to learn control, but they all get there in the end. Until that time comes, ensure that your puppy is always left with polythene on the floor covered with newspaper. You will probably find that messing on the paper becomes less and less frequent as your pup gets older and develops more control. He should be completely clean by the time he is seven or eight months old.

If your puppy reaches this age and is still not clean at night, it could be that he has developed a habit of going to the toilet at a particular time of night. This can be cured by taking your puppy and his bed up to your bedroom and confining him to his bed so that he cannot get out. When he wakes up at his usual time and finds he cannot get out, he will whine and wake you up. Take him into the garden and allow him to relieve himself.

The next night, wait for ten minutes before letting him out. The following night he should wake up slightly later; again, wait ten minutes before getting up. Eventually he will have retrained his body to last all night without the need to go out (this does not take as long as you might imagine) and, once you have established the habit of being clean all night, he can then be returned to the kitchen. Make sure you have thoroughly cleaned the kitchen floor (see page 72) before you do so.

Punishment – why it does not work

Never, ever punish your puppy if you catch him going to the toilet in the house as the distress this causes will make it difficult for him to learn what to do instead. He may begin to avoid going to the toilet in front of you because he knows it makes you angry, and sneak away to do it, making it harder for you to teach him the correct behaviour.

Similarly, it is not useful to punish your puppy if you find a mess on the floor that was done earlier. He will not learn from this – not because he cannot remember what he has done, but because he cannot relate the punishment he is receiving to the earlier act

of going to the toilet on the floor. As soon as you begin to look angry, your puppy will display a submissive response in order to appease you and turn off your anger. Unfortunately for puppies, the submissive response looks to us like guilty behaviour, so we are more inclined to punish, thinking to ourselves, 'He knows he has done wrong, because he is looking guilty.'

Not only is punishing after the event ineffective, but it may also be counter-productive. If your puppy lives in fear of you suddenly becoming aggressive for no apparent reason, he will be insecure and unhappy, which may inhibit his learning ability. Worse still, if you punish him when returning to the room, he may become anxious about being left alone, which may cause him to show all sorts of unwanted behaviour when he is older.

How long will it take?

Different puppies learn at different rates. Some pick up what is required almost instantly, while others may take much longer. Some take as long as six months or more. A puppy that came from a dirty or cramped kennel is likely to take longer than one that had a better start. Bright puppies will learn more quickly than less intelligent ones.

The biggest influence on how quickly a puppy becomes housetrained is how much time and effort you put in. More input from you will speed up the time taken to become completely clean; less input will prolong the process.

Toilet training on voice-cue

Toilet training should not end with housetraining. About seven million dogs live in Britain; many of them live in towns and built-up areas. As pressure on the available space becomes greater, the public are becoming increasingly conscious of the problems of dog fouling, especially on pavements and areas where children play.

If you want to avoid the fairly unpleasant, but necessary task of picking up after your dog in the

Many puppies are not completely clean until they are six months old so keep your expectations low and maintain the regime.

Once you begin to take your puppy out for walks, the housetraining process will become easier.

street, it makes sense to train him to go before you leave home. This is not as difficult as it may sound, but requires a fair amount of time and patience in the early stages.

If you have been successfully working at the housetraining process, by the time you are able to take your fully vaccinated puppy out, you will have a particular phrase that he will associate with going to the toilet. You should also have a fairly regular routine and will have some idea of when your puppy needs to go. Try to arrange your first walk to coincide with this time. Go out to the garden as usual, repeating your chosen phrase until your puppy does what is required, then praise him enthusiastically and take him out for a walk.

If he does not go to the toilet, take him back inside for a while and try again later. If you take your puppy out for a walk only after he has been to the toilet, he

will eventually begin to realize that producing the required deposit results in a walk.

This process takes patience to begin with, but your time and effort will be rewarded when you see other people picking up their dog's mess in the street, or when you step in some that an irresponsible owner has left behind and realize how easy your dog is to live with.

Living in a flat with no garden

Housetraining a puppy in a flat that has no garden is more difficult, but is not impossible. An area that your puppy can use as a toilet area will need to be found close to the flat. Since this may be a considerable distance for your puppy to walk, it becomes even more important that you take him out every hour without fail, carrying him if necessary for part of the way, to enable him to go outside. You will also need to be extra-vigilant to notice well in advance the signs of your puppy wanting to relieve himself.

An alternative is to have polythene and newspaper just outside the door and train your puppy to go there. However, this has the disadvantage that you will need to train him to go somewhere else later on once he has more control over his body.

Occasional lapses

It is not uncommon for housetraining to break down when the puppy is put under stress of any kind. This includes being punished unpredictably, if he is unwell or if there is a sudden change in family stability, such as Christmas, quarrels, the arrival of a new baby or any sudden change of attitude towards the puppy, for whatever reason.

Bodily changes that occur when reaching sexual maturity, as a female pup prepares for her first season or a male pup begins to lift his leg, can also cause a short lapse in the ability to be clean. Until puppies are one year old, do not expect too much. Occasional lapses during the first year are understandable and to be expected, since your puppy is still a young animal. When they occur, go back to the original housetraining programme again. As your puppy matures both physically and mentally, lapses should become less and less frequent.

Submissive urination

Some puppies, especially females, are prone to leaking urine when excited or stressed, or during encounters with high-ranking humans or dogs. They will sometimes produce a small puddle when being greeted or scolded, possibly turning on their backs or sitting with one hind leg raised. This is a natural response designed to appease and to turn off any aggression directed at them by higher ranking animals. It is their way of saying, 'Look [or rather, smell], I'm still little, please don't hurt me!'

If you shout or get cross when you see this happening, your puppy will do it more, in order to appease you more. The best response is to ignore him and walk away until you get to a place where any leakage will not matter. If your puppy makes a habit of greeting people at the door in this way, keep a sheet of polythene and an old towel close by and manoeuvre him onto this before greeting him or allowing him to be greeted by visitors.

Puppies that do this need to have their confidence built up. This will happen gradually as they get older, but try to avoid telling them off or being cross with them. Show them what you want them to do instead. Such puppies are usually very eager to please and will do what you want as soon as they know what it is.

Puppies that roll over when being greeted are prone to developing submissive urination problems if scolded or punished.

CHAPTER 8
Behaviour control and leadership

Puppies do not arrive with an inbuilt set of rules about how to behave well in our world. We need to teach them patiently how to conduct themselves appropriately. This may appear to be a daunting task, but it is easy once you know how. It is important to encourage good behaviour and reward it immediately, as well as making sure that your puppy's needs are met. Once the puppy is content and knows what will be rewarded, it is relatively easy to prevent and stop unwanted behaviour. Being a good leader and earning enough respect to ensure that your puppy is responsive to you will also help with this process.

Encourage good behaviour, prevent bad behaviour

Throughout your puppy's first year encourage him, whenever possible, to show the correct response. Praise and reward immediately and these responses will soon become good habits.

Preventing unwanted behaviour means that you need to think ahead. Try to anticipate what your puppy might do, distract him from it or physically prevent it from happening. Encourage the behaviour you do want instead, then praise and reward well when your puppy shows the correct response.

Think ahead and prevent unwanted behaviour, using a lead if necessary. Make sure all rewards come from you, rather than through the puppy's bad behaviour.

It is always better to stop unwanted behaviour before your puppy realizes how good it feels to do it. Otherwise, the pleasurable feeling will encourage him to do it again and it will quickly become a bad habit. If your puppy is prevented from engaging in and practising an action until he is mature, he is unlikely to attempt it when older. This applies to all kinds of unwanted behaviour, from jumping up to getting on the furniture to chasing livestock or joggers.

If your puppy does something you do not like, stop him at once by taking his collar and gently leading him away from temptation, showing what you want him to do instead. If he is repeatedly doing something unacceptable, leave a trailing lead or house-line (a line about 2 metres/6½ feet long) attached to his collar to enable you to get hold of him more quickly, without grabbing for him or chasing him around. Be careful that children or frail adults do not trip over the line, and only leave it on when you are there to supervise.

Meeting your puppy's needs

Your puppy will be much more pleasant to live with, better behaved and less 'naughty' if you try to ensure that all his behavioural needs are met. A puppy's body is constantly trying to rebalance itself and make sure it has everything it needs for maintenance and development. Puppies behave in ways that try to make up for any internal deficiencies. Ensuring that their needs are adequately met before they become desperate prevents much unwanted behaviour. Puppy behavioural needs fall into three categories: safety, social and body maintenance.

Puppies need to know that their owners will keep them safe and protected.

Safety needs

Puppies need to feel safe and to rely on their attachment figures (their mother or new owners) to help them out if they are in danger. Ensuring that your puppy feels safe and protected is important because this prevents him learning to be aggressive and showing other unwanted behaviours that are associated with fear and anxiety.

Social needs

Social attachments are very important to young puppies. In a human family, they will be most content if they have one or two main attachment figures with whom they can make strong and lasting bonds. These people need to have enough time to give the puppy plenty of love and attention regularly throughout the day. If a puppy's social needs are not adequately met, he is likely to show unwanted attention-seeking behaviour or frantic over-activity when people are present, or he may become withdrawn and asocial.

Key points

- Encourage and reward good behaviour.
- Think ahead and prevent unwanted behaviour.
- Stop unwanted behaviour at once.

It is important for young puppies to be given a quiet place to rest and get plenty of sleep.

Body-maintenance needs

Puppies need to sleep, play, exercise, explore, eat, drink, stay warm and chew. Failure to provide the opportunity for your puppy to do these activities on a regular basis will result in unwanted behaviour. This is because the puppy has an internal drive to do these things and will try to do them, whether you like it or not.

Failure to give a variety of suitable chews, for example, will result in unwanted chewing of household items. Lack of quiet resting places to sleep without disturbance (common in busy households with children) can result in an over-tired, irritable puppy. Even something as simple as feeding many small meals a day so that the puppy is not ravenous (and hence 'difficult' and irritable) for a few hours

before the only meal of the day can make a big difference to its behaviour. And, most common of all, puppies denied adequate opportunities to play, exercise and explore are constantly trying to do this in the house, getting into trouble for biting at plants, grabbing at children, boisterously jumping up or getting over-excited at inappropriate moments. Ensuring all of a puppy's body-maintenance needs are met is very important and will result in a contented, placid puppy that is easier to raise and train.

Sleep

Ensure your puppy gets enough 'quiet' times to rest. This is particularly important if you have a busy household. Give your puppy regular 'time-outs' in a quiet, peaceful place with a comfortable bed. Make sure children understand that it is important to leave the puppy alone when he is asleep, and that they must not wake him up.

Play

Make time to play with your puppy for a short period every few hours throughout the day. Some puppies need more play than others, so try to play just enough for your puppy to lie down contentedly in between play times (see page 92).

Exercise

Different puppies require different amounts of exercise. It is not possible to take puppies on long walks because this will damage their growing bones and joints. Instead, provide plenty of opportunities to run free in enclosed areas and play active games with toys, and mentally exercise them by giving plenty of new experiences and socialization. If your puppy is very boisterous, he may need more exercise than you are currently giving in order for him to be relaxed and

comfortable. This is particularly the case if you are out at work for long hours. You will need to spend more time keeping him active and using up his energy when you are at home.

Leadership and boundary-setting

As your puppy grows older and begins to understand what is required of him, there will be times when what you want him to do is not compatible with what *he* wants to do. As he gets older, it will become harder to distract him and refocus him on doing what you want him to do instead. Gradually it will become more important that you are respected as the leader. Along with learning to accept your authority, your puppy also needs to learn to deal with the frustration he feels when he cannot get his own way.

Why being the leader is important

A dog that views his own wishes as less important that those of his humans will be much easier to live with, better behaved and more willing to obey voice-

Providing enough exercise for your breed of puppy is essential if he is to be calm and well behaved.

cues. A well-behaved, obedient and compliant dog is a pleasure to own and is generally much happier than one that is in constant conflict with his owners.

In addition to having a dog that is better behaved, it is much easier to train a dog that views your requests as important. If a dog feels subordinate to you, he will try to please you and will learn from you, in much the same way as children will learn from a teacher who maintains a certain air of authority and who commands respect. Training a dog that thinks he has the right to make decisions is nearly impossible.

A dog that thinks he is pack leader will be making his own decisions. He will be out of your control and will act like a spoilt child, trying all manner of tricks to get his own way. He will not obey requests or do anything he does not want to, and he will always want to be the centre of attention. In our society no one can afford to have a dog that is out of control, and dogs that think they are superior to humans often have a very short life.

How to be a good leader

Earning the position of leader is easy with a puppy, because he will automatically look to you for guidance. Keeping that position as your puppy grows is important, particularly during adolescence, when he gets bolder and will begin to test the boundaries. With most domestic dogs this is not difficult, since they have been bred for generations to be biddable and to work with humans. Compared with their wild ancestors, the wolves, our pet dogs care very little about hierarchy, preferring instead to cooperate and

You cannot force your puppy to respect you. Respect needs to be earned through well-thought-out training.

A relationship based on trust and respect is an essential foundation for good behaviour.

live in harmony with us. However, some dogs are more strong-willed than others, and all benefit from viewing humans as superior since they need to learn to fit into our world, rather than us living around them.

A good leader does not bully or force others to obey. Instead, a good leader earns respect by being smarter, braver and more decisive than others, winning challenges and contests and demonstrating that it would be better for all if they made the decisions. In a dog pack, it is not the largest and strongest that is top of the hierarchy, but usually the one with most determination, the strongest will and the one who will risk everything to win.

The best pack leaders are usually benevolent and tolerant, but can be tough and uncompromising whenever they need to be. They will stand no nonsense, but are happy to be their dog's friend. A good pack leader knows when to stop, and does not constantly bully the dog to force him to stay inferior. Good pack leaders have four main attributes. They are:

• Able to keep the pack safe
• In control of resources
• Good communicators
• Able to win contests and challenges.

Controlling resources and your puppy's movement around the home will teach him to be patient and respectful.

Keeping the pack safe

Leaders need to be brave and smart enough to keep their pack members out of trouble and solve problems quickly and efficiently if any are in danger. Young puppies will see danger in a variety of situations, particularly if they are not well socialized. Watching your puppy and making sure he is comfortable and feeling safe in all situations – and doing something about it if not – is essential if you are to be seen as a good leader (see page 54 on socialization and page 114 on preventing aggression).

Controlling resources

If you have control of the resources, you will be making important decisions that affect the puppy's life, such as how often and how much he eats, when he plays, how much exercise he gets and how much access to the outside world he has. Making your puppy work for what he gets by complying with your requests, as well as deciding when and where he should get access to these important resources, will put you in control of your puppy's life and actions. This will be observed by your puppy, and his view of your status will increase. Giving resources too freely or on demand will allow your puppy to think he is in control instead.

You must decide when your puppy should sleep, eat, play and have social contact. If you control the resources, you are more likely to be respected as leader. And since you are in command of all the resources, it is important to make sure that all your puppy's needs are met in time to prevent him from behaving badly in order to try to get them himself.

Once your puppy is trained, ask him to come to you, or sit, or lie down, or wait for the resources you are offering. Asking for compliance regularly, and rewarding him well, will make it more likely that your puppy will view you as a leader and respond automatically in future.

Communicating well
Good leaders are able to communicate well with others. Your puppy has not evolved to be able to understand spoken words easily, but will be more

> ### Resources you control
> - Food – dinner, treats
> - Play – games with toys
> - Social contact – praise, fuss, gentle stroking
> - Access – to outdoors or places to explore and walks.

responsive to gestures and body language. Being a patient, careful teacher, who takes care to communicate clearly until an understanding is reached, will earn you more respect than shouting unknown verbal commands that frighten and confuse.

Winning contests and challenges
In the wild, wolf cubs or wild dog puppies will play-fight among themselves and among older members

Keeping control of games and winning more often than you lose will teach your puppy self-control.

of the pack as they are growing up. During these contests they learn about their own strengths and weaknesses relative to other pack members and, hence, where their natural position in the pack lies. In the same way, our puppies will play with us and learn similar lessons.

Playing to win

Strong-willed dogs (and people) play possessive games and play to win. If your puppy continually initiates and wins games, it may help him decide that he is stronger than you, both physically and mentally. To help you become a respected leader, particularly if you have a strong-willed puppy, keep to the following rules during games, especially those of tug-of-war:

- Play with him with toys that you keep in your possession out of your puppy's reach. Give him toys that he can play with by himself, but do not get involved in games with them.
- Invite your puppy to play often. Since you have the toys, it is your responsibility to remember to play.
- When playing tug-of-war games, win more often than you lose. If you always win, your puppy will lose interest in the game (no dog or human enjoys games they are not very good at). By winning more often than you lose, your puppy will enjoy playing and will be learning that you are a more determined opponent. Teach him to stop tugging on the toy and to let go immediately when asked (see page 106).
- If necessary, use a house-line to help bring your puppy to you so that you can get the toy back easily; this 2 metre (6½ foot) line trails around after your puppy and removes the need to chase him to catch him so that you can retrieve the toy.
- Play until you have had enough, and try to stop before your puppy gets bored.
- Take the toy that you have been playing with away with you at the end of the game.

Winning challenges

If your puppy is very strong-willed and more intelligent than the average puppy, he will find all manner of ways to challenge your leadership, particularly when he is becoming bolder during adolescence. Provided you are aware of this and take care to maintain your leadership status in the ways described in this chapter, you should find it relatively easy to stay in control.

If you find that there are some situations in which he is behaving badly, think about how he may be winning during encounters, why you are losing and why you are unable to stop him. Then work out a way to overcome the problem so that the situation is reversed and you become the winner. You may need help from a pet-behaviour counsellor (see page 203 for contact details) and, if you are having problems, it is better to ask for advice sooner rather than later.

How ambitious is your puppy?

All puppies are different. Both their genetic make-up and their early experiences will play a part in determining how ambitious they are. Some can be given all the privileges usually reserved for pack leaders and still not take advantage. Others will want to be in control at all cost and will exploit every opportunity to do so. If your puppy is naturally ambitious, you will need to win more games and challenges and be more controlling of resources in order to stay in command. More submissive puppies can be allowed to get away with much more.

You need to assess your puppy as he grows up and adjust his achievements accordingly. A subservient puppy can be made more confident by arranging for him to win more often, whereas a very pushy puppy may need to lose many encounters with you if he is to respect you as leader. The important thing is to maintain a balance so that your puppy unquestioningly considers humans to be higher in status, but not so high above him that his true character is inhibited.

The 'Off' voice-cue

It is very useful to teach your puppy a voice-cue that means 'Take yourself away from what you are doing and you will be rewarded.' This can then be used for times when your puppy is about to do something unwanted – for example, when he is too near food on

Establishing ground rules, such as not getting on the sofa, early in your puppy's life helps to avoid problems developing in adolescence.

a low table, is approaching someone who doesn't like dogs or is trying to pick up something you don't want him to have.

Repeat the 'Off' voice cue exercise over many small sessions until your puppy will move his nose away instantly when you say 'Off'. Then slowly build up the time he waits before you give the treat, counting up to ten eventually, to help him learn self-control and good manners.

Place tasty food items on a low table, say 'Off' and reward him well, away from the table, with an even better treat for complying with your request. Then think of other situations and places where you may need to use the 'Off' voice-cue and teach again in a similar way, gradually building up his experience until he learns to respond anywhere.

When to say 'No'

At some point in their lives, most puppies have to learn that they risk unpleasant consequences if they continue to do something that you have warned them not to do. Only rarely should you need to tell your puppy off. Most of your interactions should be happy and pleasant ones. This will ensure that you become a good friend and that your puppy will try hard to please you.

Exercise: The 'Off' voice cue

▷ **1** Begin by offering and feeding your puppy a tasty treat. Offer another treat, but curl it into your hand so that he cannot get it. Say 'Off' calmly and clearly (do not shout it aggressively!). Keep your hand very still and ignore any of his attempts to get it. Hold it high enough so that he cannot use his paws, and do not respond to any nips or chewing (if you cannot cope with this, wear an old leather glove to protect your hand during the first phases of training).

There are two types of situations where discipline may be needed. Firstly, if your puppy does something totally unacceptable, such as jumping up at a child, it is better to correct this behaviour immediately than let your puppy be rewarded by it, and therefore want to repeat it in future. It is best to save correction for behaviours that are highly unacceptable. For milder problems it is better to restrain your puppy, teach an appropriate behaviour and reward him for doing the right thing instead.

Secondly, a bold puppy (especially during adolescence) may be ignoring your voice-cue deliberately because he would rather do something else. Immediate correction will teach him that he must respond to you when you give him important commands. Before using correction in this situation, you must be absolutely sure that your puppy not only fully understands what to do, but has also heard your voice-cue.

How to do it

If a reprimand is needed, it should be immediate, startling, effective and over in seconds. Your correction should be followed immediately by showing him the right behaviour and rewarding him for doing it.

△ **2** Wait patiently and without saying anything. Eventually, your puppy will give up and move his nose and paws away.

▷ **3** *As soon as* there is a small gap between your puppy's nose and your hand, open your hand to release the treat. Timing is very important – try to make your response instant, as soon as he moves his nose away.

If your puppy is about to do something you don't want, try to stop him by saying 'Off' or giving another appropriate voice-cue such as 'Sit', which will prevent him doing the unwanted behaviour. If he is bold enough to ignore you and continue with what he planned anyway, you need to follow up immediately with a correction. If he has already done something unacceptable, use the correction immediately.

The effectiveness of a correction lies in the element of surprise. You will need to shout 'No!' or 'Aargh' with enough decibels to startle your puppy and prevent his intended behaviour. Block his route to whatever he wanted to do with your body if you can, and physically stop him if you need to. Then use eye contact and your voice to get him to focus on you (this is a lot easier if you have already earned the right to be leader, see earlier). The correction should be used very infrequently. Most of your communication should be in a calm, quiet voice so that turning up the volume suddenly is immediately effective.

After an effective correction, your puppy's attention will be riveted on you. It is important that you use this time to show him the required behaviour in this situation and praise him for doing the right thing. In order to do this, you need to revert to being calm and pleasant so that your puppy is no longer worried by you and can concentrate on what you require. This will allow him to learn that it was the particular behaviour that made you act in such a way, and he will be less likely to do it next time.

Amount of correction needed

All puppies are different, and they all have different levels of sensitivity. For some puppies a slightly raised voice is enough, whereas with others more volume is needed to let them know they are wrong. A correction that may seem mild to a bold puppy may feel like the end of the world to one that is more sensitive.

Why physical punishment does not work

It is not necessary or desirable to use physical punishment in order to bring up a well-adjusted, well-mannered adult dog. Not only is it unnecessary, but it is very unpleasant for the puppy and an excess of it can cause resentment and fear of you and a mistrust of humans in general. It will weaken the bond between you and your puppy and spoil any friendship and trust between you. Families who have used physical punishment to discipline their dogs often have aggressive dogs. Dogs that come from a family where no aggression has been used to raise them are rarely aggressive themselves.

If hands are used to administer the punishment, the puppy may learn to be hand-shy. Later, if he sees a hand coming towards him – for example, if a child reaches out to pat him on the head – he may bite in self-defence. If the puppy learns that human hands never hurt and that they sometimes bring rewards, there will be no need for him to bite hands that are reaching towards him.

Unfortunately, physical punishment seems to be the first resort for many people when things go wrong. It is much easier to punish than to take considered action, and many gadgets are now being sold for this purpose. Sadly, punishment sometimes appears to work well and to work quickly, thereby encouraging owners to apply it more often. It appears to work so well because the symptoms of the problem are being treated, rather than the root cause.

For example, take the dog that is afraid of strangers because of a lack of socialization and that growls whenever they approach. If his owner applies sufficient punishment whenever this happens, the dog soon learns not to growl because he becomes too afraid of the consequences. The owner will think his treatment has worked and will recommend it to his friends. However, the real cause of the problem – the fear of strangers – has not been treated and when, one day, the dog feels seriously threatened, he will bite without warning.

Ignore mistakes
- Never use any correction while training your puppy. Ignore any mistakes he may make while attempting the exercises, and instead reward the behaviours you do want.

A loving, trusting relationship with your puppy will lay the foundation for good behaviour as he matures.

Understanding is the key to effective cures for problem behaviour. Punishment may seem like an effective way out, but it rarely is. If you are having problems with your puppy's behaviour, try to think about things from his point of view. This will often lead you to a much better solution than one arrived at merely because it makes life easier for you. We need to use our intelligence to solve problems, rather than resorting to aggression in an attempt to try to force our dogs to behave.

Many dogs are punished for not understanding what their owners want them to do. Training is a long process, and few owners take the time and trouble needed to produce a well-trained dog. Many people do not understand that it takes many repetitions of the exercise for dogs to learn voice-cues, and as a result they expect too much of them. Punishment will not help a dog to learn more quickly.

Behaviour problems

If, as your puppy matures, he develops behaviour problems that you cannot understand or deal with, seek help from someone who has been specially trained to deal with such problems. Your vet should be able to refer you to a dog-behaviour counsellor who is a member of the Association of Pet Behaviour Counsellors (see page 203 for details).

Do not despair if things go wrong. Owners frequently have difficulties with their dog's behaviour, especially in the early years, and it is no disgrace to admit it. It is always best to seek help and good advice sooner rather than later.

CHAPTER 9
Toys and games

How we play with our puppies is far more important than many people realize. This chapter will help to ensure that your puppy learns only good things from play sessions, and that he learns to channel his energies into fun games with humans rather than using them up in inappropriate ways.

Why play is so important

Puppies that learn to play human games grow into dogs that see humans as a source of pleasure and enjoyment. Such dogs are much more fun to have around and are more sociable. They probably have a better life in human society than those that have only learned to play with other dogs.

Playing games is more than just a way for you and your puppy to have fun, although having fun during games is essential. Games enable the participants to find out about each other and learn about the qualities, traits, strengths and weaknesses of the other player. Fear of the unknown reduces during play as the puppy and owner become more familiar with each other. New things are discovered as both seek ways to enhance the game, and a better understanding will develop between the players.

Playing regularly with a puppy develops a strong bond between you both. It is possible to see a dog every day for years and even be his only source of food without developing a strong bond with him.

Playing games with your puppy is fun and a way of building a strong bond and developing trust between the two of you.

Puppies are naturally playful and it's important to try to channel this energy into games with toys.

Several play sessions, however, will help you form an attachment and will provide a way of establishing friendship and mutual trust. Being friends with your puppy is essential if he is to be trained easily and effectively. If he really wants to please you, and you are fun to be with, he will work harder to do as you ask.

The more games you play with your puppy, the more he will consider you to be the most interesting thing in his world. The more he wants to be with you and please you, the easier it will be to control him.

If your puppy stays near you on a walk, for example, waiting for you to throw a toy, he will not wander off by himself and get into trouble. If he wants to chase the toy you are carrying, he will not find his own games by chasing cats, joggers, livestock or cyclists. As you are so interesting, he is likely to rush

back to you in case you are going to play a game with him, and he will not continue to run around with other dogs in the park, making you late for work while you try to catch him.

Games are also an outlet for dogs' natural hunting abilities, many of which they retain. Directing your dog's instinctive desire to hunt, chase and kill prey onto toys will prevent him from finding outlets for these activities elsewhere. Some dogs, particularly those of the working breeds, are bred to be physically active and mentally alert all day. This is just not possible in most pet homes and, in the absence of any work to do, play is a useful substitute.

Adequate physical and mental exercise is essential, especially as your puppy matures and becomes a young adult. Lack of exercise can result in a bored,

Your puppy's breed will dictate his favourite game. Breeds like this Staffordshire Bull Terrier prefer possession tug games.

discontented dog with more energy than he knows what to do with. Very often, such excesses of energy are the cause of unwanted behaviour as the dog finds alternative outlets for them. Several strenuous play sessions, given at varying intervals throughout the day, can result in a dog that is well adjusted, contented and looking forward to the next time you bring out the toys.

Since puppies (particularly those of the large, heavy breeds) should not be walked long distances until their bones have matured, play with toys provides a good way to use up some of their exuberance. Several short play sessions spread throughout the day are preferable to one long play session. Try to play with your puppy when he is being good – that is, when he is not doing something you

disapprove of. You will then be rewarding this good behaviour and he will be more likely to do it next time. If you only get the toys out when he is becoming a nuisance, his general behaviour will get worse.

Once your puppy knows just how much fun games with toys can be, you can use the toys themselves as a reward for a training exercise. Games with toys can also be used to your advantage in other ways – for example, overcoming your puppy's shyness with other people or any other fears that he may develop (see page 68).

Different games for different dogs

A toy can be anything that is non-toxic and will not splinter or cause harm in any way: for instance, an old sock that has been stuffed with other old socks, or a

Terriers often enjoy games with squeaky toys as they make a sound like captured prey.

piece of strong fur fabric or sheepskin. Toys should always be large enough that they cannot be swallowed, and should be removed before they break up into smaller pieces.

Small puppies prefer soft things that are similar to their litter-mates, so buy large, fluffy soft toys from pet stores or charity shops. These will usually be preferred until the puppy becomes an adolescent, when he will gradually progress to harder, more solid toys that can be thrown further.

The various breeds of dog have been bred to serve different purposes and, since games are a substitute for the work they used to perform, different breeds will prefer to play different types of games. There are basically three types of games:

- **Chase-and-retrieve games** – often preferred by herding dogs, gundogs and hounds
- **Possession games (tug-of-war)** – generally preferred by guarding and bull breeds
- **'Shake-and-kill' games (squeaky-toy games)** – often preferred by terriers.

Most puppies will play all games, but will often prefer one type to the others. As your puppy begins to get older and show a preference, it is advisable to buy toys to suit the type of game you are playing. It is, for example, much easier to play games of chase with balls and Frisbees than to throw a knotted rope. Similarly, it is easier to remove a rubber pull-toy from the mouth of a tug-of-war player than it is to try to fish out a ball.

Ringing the changes with toys will help to keep both you and your puppy interested. Puppies are no different from children in this respect, and will appreciate a new toy or a different game occasionally. Since you are also a participant in the game, you need to stay interested in the game too, and sometimes a new toy (or rediscovering an old one) can improve games for both of you.

Keep a toy in your coat pocket so that you always have one to play with when out on a walk. Playing with toys instead of sticks will prevent injuries to your puppy's mouth, eyes or face. These often happen when one end of a stick becomes stuck in the ground, the puppy cannot stop in time and runs into the other end. Simply ignore your puppy when he picks up a stick and he will soon learn that it is more fun to play with the toys you are carrying.

It's a good idea to keep a toy in each room so that they are always available. You only need a short amount of time to play a game, and if toys are readily available you can play whenever you have a few minutes to spare.

Games with toys v. fighting games

For all young animals, play is an important way of learning skills and strategies. Puppies learn a great deal in this way, so it is essential that they learn the right things. The games you teach, the rules that are applied and the skills they learn during these games will affect the way they behave later on.

For this reason, puppies should be taught how to play with toys rather than being allowed to continue with their natural game of play-biting. A puppy that is encouraged to play rough-and-tumble games with humans, to bite at our arms and legs, becomes well practised in these skills. Even dogs with good temperaments may bite in certain circumstances (for example, if they are injured in an accident and in great pain), and people are less likely to be bitten badly if the dog has no clear idea how to go about it.

Skills that are learned in puppyhood are put into practice later on. If you do not want your dog to know how to bite humans, should he decide there is a need to do so, do not teach him how to do this now. He will call on what he learns in play, should he need to react quickly in times of stress or excitement. Teaching him that there is no need to bite is important too (see page 114), but if you do not want a dog that knows how to bite people, do not teach him; orientate all his biting behaviour onto toys instead.

Play-biting

When your puppy first arrives at your home, he will have been used to playing with his brothers and sisters. Play-biting is the natural game for puppies and, when he has settled in to your house, he will attempt to play with you in this way too. You will need to teach him how to play with you using toys instead (see page 114 for dealing with excessive play-biting).

To do this, have a large soft toy ready every time you choose to spend some time with him during the

Use a large soft toy to entice your puppy to play and keep it moving so it is more exciting to chase and catch.

early weeks. As soon as he gets excited, he will probably begin to chew on your hands or bite at your fingers. Make a fist with the hand he is chewing to make this more difficult, but keep it still and distract him at once by offering a toy instead. Make the game interesting by keeping the toy moving, wiggling it or rolling it along the ground in front of him (if you throw it, he will have no idea where it has gone since young puppies have to learn this manoeuvre). At first he may try to chew on your moving hands instead of the toy, but persevere and he will soon learn that toys are much more fun.

As your puppy learns to play games with toys, make it a rule that any unacceptable behaviour

results in an end to the game. Behaviour that is unacceptable includes anything that will be unwanted as he becomes an adult. Think of him when he is full grown playing with a child. Would it be acceptable for him to growl at the child, even in play? Would the child be upset if your dog worked his way round the toy during a game of tug-of-war and bit their fingers to make them let go? Would it be all right if the dog came racing towards the child and tried to snatch the toy out of their hand?

Remember, anything you teach now will become your dog's normal behaviour later on. Stop any unacceptable behaviour by quietly ending the game and walking away with the toy as soon as it occurs.

Your puppy will soon learn what causes the game to end, and will begin to avoid such behaviour so that the game can continue.

Keeping certain toys to yourself

It is essential that your puppy has a variety of toys and chews to play with and to amuse himself with at all times, as this will help prevent him from chewing and playing with valuable household objects. Do not get involved in games with these toys, but have another set that are kept in your possession. These toys will be more interesting to your puppy than those kept with him and because he associates them with having fun with you, they will mean much more to him and he will work harder for them.

Such toys can also be more delicate than those that are left with your puppy all the time. Squeaky toys, for example, will last longer if played with only when you are present; they would be chewed up quickly if left with a puppy going through the teething stage. Owners will often say their dog has no toys because as soon as they give him a new one he buries it, chews it to pieces or hides it. By giving him indestructible toys to play with by himself and

Exercise: The retrieve

◁ **1** Begin by teasing the puppy with the toy and rolling it gently along the ground. As your puppy starts to run after it, give a voice-cue (such as 'Fetch it!') so that he begins to associate this word with running out to pick things up.

▽ **2** Once he has picked up the toy, use your voice to praise him and encourage him to come back to you. If you sound exciting enough, he will probably trot back to you carrying the toy in this mouth and wagging his tail. Keep your hands still, so that he does not think you are waiting to take the toy from him, or he may try to avoid you and lie down with the toy elsewhere.

keeping delicate toys to yourself, you will always have certain toys with which to play with your puppy, and the toys themselves stay novel, interesting and safe.

Teaching the retrieve

This is a really useful exercise since it allows you to get toys back easily and to stay in control of games. Playing 'fetch' games is an easy way to exercise your puppy, and the retrieve represents the basis of many other games that you can play with him as well as useful 'tricks' to teach him, such as fetching the post or helping you bring the shopping in. Teaching a puppy to retrieve is easy if you start early enough, before he has learned to avoid people once he has the toy.

Children and games

Children and puppies usually play happily together, both getting a lot of fun from the experience and learning a great deal in the process. However, puppies can just as easily learn the wrong ways as the right ones, and it is essential that an adult is present during play sessions to supervise the proceedings discreetly.

▷ **3** When he reaches you, do not touch the toy, but make a big fuss of him, touching him anywhere except around the head region. If you go straight for the toy, he will learn to avoid you and will eventually begin to stop coming back. When he has had the toy for a few minutes (be patient!), he may drop it of his own accord or may settle down to chew it. Take hold of the toy and wait until he lets go. As he does so, give a voice-cue (such as 'Drop') and remove the toy. Praise him and throw the toy for him to chase again. As your puppy grows older and gets into the habit of coming back to you, you can begin to take the toy away more quickly (see page 106 for getting him to 'Drop' on a voice-cue).

Supervise children when they are playing games with your puppy to ensure both are learning acceptable behaviour.

Remember that anything your puppy learns to do when young is the pattern for his future behaviour. Children would not wish their puppy any harm, but they do not always realize the consequences of their actions. Observe the games that your puppy plays with children. If he begins to play unsuitable games, either correct him or encourage the children to play a different game with him.

Keep games fun

Puppies are naturally playful and it is not difficult to encourage them to play with you from the outset. The only exception may be a very shy puppy, which may need time and careful coaxing before he feels bold enough to attempt a game.

What you do during games will have a great effect on whether your puppy considers you to be an exciting playmate or not. Children are usually best at games, since they are often less inhibited than adults. Imitating their ways by squeaking, moving quickly, laughing and being excited makes a puppy excited, too. Be as silly as you like!

If you are a fairly reserved person, unused to being silly in public, practise in private. It will become second nature and you will find yourself playing daft games with your puppy in company as well. Onlookers will not think anything of it; everyone expects owners to be silly with their dogs, especially when they are puppies.

Movement excites puppies, so keep the toy moving erratically. Tying your puppy's toy to a piece of string and using it to make the toy 'jump' may encourage a reluctant puppy or revive an old game. Be inventive and try new games whenever either of you begins to get bored with old ones. Make games fun for yourself too. Nothing motivates a puppy quite like our smiles and laughter.

Competition encourages some puppies to try harder to win the toy. If you throw a toy and allow your puppy to fetch it, he will enjoy the game of chase. However, if you throw it and sometimes race after it yourself, getting to it just before your puppy does and snatching it away before he reaches it, or sometimes allowing him to get to it first, he will put much more effort in next time and will have more fun doing so. Teasing your puppy with the toy before you throw it will have a similar effect, but do not overdo

this and do not encourage children to do it as they will often do it to excess.

Exciting times of day

Puppies will want to play games whenever they get excited about something. This could be at any time, from you coming home to visitors arriving. During these exciting times of the day you will often see dogs playing their favourite games. Labradors may carry a slipper when you come home, terriers may grab at and 'kill' papers put through the letter box, and some dogs may play tug-of-war with the lead during the initial excitement of getting ready for a walk. You can

Puppies will try to play their favourite game, like this one holding a slipper, whenever they get excited.

use this desire to encourage a reluctant puppy to play with you by getting out toys whenever he is excited about something.

For example, keep a toy by the door and get it down to play with when you arrive home. If you link all the exciting things of the day with the appearance of the toy, eventually its appearance will become exciting. Since dogs like to play their favourite game when they get excited, your puppy will eventually want to play with the toy whenever it is produced. This will have the added advantage of preventing him from finding other, inappropriate games to play during times of excitement.

At first, do not try to control the games too much or put in too many rules. Nothing puts a puppy off playing more than playing with someone who is being bossy or who always takes control. Until much later on, do not insist on perfect retrieves or a sit-stay (see page 186) before the puppy is allowed to chase after the toy. Get the enthusiasm for the game first, and put in the control afterwards.

Try not to get angry during games, which can sometimes happen if they are not going the way you intended. Stop the game and take time to think about what went wrong and how to prevent it next time, rather than entering into a confrontation with your puppy. Do not force yourself to play when you really do not feel like it or are over-tired. If you become nasty during games, you will put your puppy off and teach him that you are not to be trusted. Always finish the games on a high note, before either you or your puppy tires of the game.

Games should be an entertainment and a recreation for both of you. If you can make all games fun, you and your puppy will have a great time and the bond between you will grow stronger.

Controlling the games

If you want an adult dog that is always under control and obedient, even during times of stress and excitement, teaching your puppy some control during games is essential. Lessons and routines learned through games will be invaluable during everyday life,

as well as in a crisis. They become particularly important if you have acquired a puppy of a large or powerful breed.

Imagine your puppy as an adult, racing after next door's cat, which is heading for a busy road; or getting loose in a field of livestock. What chance do you have of calling him back if you cannot stop him when he is in full flight after a moving ball? If you cannot get him to 'leave' a squeaky toy, how will you stop him when he is closing fast on the pet hamster, or when he has just noticed that the guinea-pig next door is squeaking? If you cannot make him let go instantly during a tug-of-war game, how will it look if he begins to play with the coat sleeve of a child who is screaming in fear, and you have to go over and wrestle with him in order to get him off?

Teaching your puppy some control voice-cues during games will enable you to control him during times of excitement or in stressful situations. Being in control of the games means that you are in control of your puppy. These control lessons should only be

Asking your puppy to sit and wait for a few seconds until the toy is thrown will teach him self-control and good manners.

Use a cord looped through the collar to hold your puppy once the ball is thrown until he relaxes and you give permission to chase.

taught once enthusiasm for the games has been established. Do not expect miracles at first. Remember that your puppy is still very young and it takes time to learn self-restraint.

Chase games

Being able to control chase games is particularly important if you have a puppy from one of the hound or herding breeds, especially if he will be large or powerful as an adult. Your puppy will need to learn two things:

- To wait while a tempting moving toy is thrown, and not go after it until he is given permission
- To come back when called, even though he is in full flight after his toy.

Waiting for permission to chase

Every so often during a play session, slip a short piece of line through your puppy's collar, hold both ends of the line and tell him to 'wait' before you throw the toy. He will attempt to move forward to chase the toy as you throw it, but will be prevented from doing so by the line holding his collar. Hold on if he struggles and wait until he is calmly accepting the restraint. Praise him gently for waiting. Then let go of one end of the line to release him, give your voice-cue to 'fetch' and encourage him to retrieve the toy.

Every so often reward him for staying with you, by keeping hold of the line and dropping his favourite toy to him while you praise him and keep him in position beside you. If you keep him guessing as to whether he has to run out to get the toy or whether the toy is coming from you, he will learn to wait for you to decide what he should do.

When he is reliably waiting when you ask him to, you no longer need to hold him, but be sure it is firmly fixed in his mind and go back to restraining him at any time if necessary. Remember that control reduces enthusiasm, so if you find that your puppy is becoming reluctant to run out, you could be overdoing this exercise and will need to allow him to chase the toy straight away more often.

Later, you can extend this exercise by making the object to be chased more interesting. Asking your puppy not to chase a familiar toy rolling along the ground is one thing; asking him not to chase a hare that he has just flushed out of long grass is another. However, it is possible to simulate this degree of excitement by offering your puppy a more exciting chase game. This will depend on what motivates your puppy, but it could perhaps be achieved by exciting him and throwing a new or favourite toy a great distance, or by seeing that he wants to join in an exciting game with other dogs or children and asking

Exercise: Elementary chase recall

▷ **1** You will need two people and two toys. Begin by throwing a toy past your partner, who should be standing facing you some distance away. Do this several times and allow your puppy to chase after the toy and bring it back to you.

him to wait as the dogs or children race by him, before releasing him to play with them. Be inventive, watch your puppy and you will soon find something exciting with which to practise before you encounter a real situation.

Elementary chase recall

This is a very useful exercise given that most dogs, at some time in their lives, will be tempted to chase something they should not. Being able to call your dog back from a moving object is something that few owners are able to do, but it is relatively simple to teach, especially if you begin with a young puppy. You need to have a reliable retrieve before attempting this, and the exercise should only be done with puppies over 20 weeks as a lot of fast running is involved which would not be suitable for developing bones and joints.

Since you are only stopping your puppy from running after the first ball once in every ten throws, you will not be able to do much of this before your puppy is tired. It may be several weeks before he begins to get the message so be patient. He will become more reliable with practice. If you shout 'Leave!' more than once in every ten throws, he will start to hesitate before running out, which will reduce the effectiveness of the exercise.

Once you have a reliable chase recall, you will need to reteach it in more exciting circumstances, as in the waiting for permission to chase exercise (see page 103), until, eventually, he will be so well trained that he will leave anything he happens to be chasing.

▷ **2** About once in every ten throws, at random, throw the toy towards your partner, who should catch it and tuck it away out of sight. As your puppy begins to run after the toy, shout 'Leave!' loudly. Since the toy he was chasing has effectively disappeared, he will eventually look back at you to see if you have it.

When this happens, tease him with the other toy and throw it in the opposite direction. Allow him to chase and retrieve this one, but try to make this chase more exciting, either by pretending to go after it yourself or by using a more interesting toy.

Eventually, your puppy will learn that when he hears the voice-cue 'Leave!', there is no point in continuing with the chase because the more rewarding chase game is now back with you.

Tug-of-war games

When playing tug-of-war games, win more often than you lose. If you always win, your puppy will rapidly lose interest in the game (no dog or human enjoys games they are not very good at). If you always let the puppy win, he will get the message that he is stronger than you. By winning more often than you lose, your puppy will enjoy playing and you will be giving him the right message about his status and strength.

Allowing him to win occasionally means having to let go of the toy at some point. If you have great difficulty in recovering the toy once you have let go of it, attach a house-line (a 2 metre/6½ foot length of cord, one end of which is fastened to his collar, the other end left loose) to his collar before doing so. If your puppy teases you with the toy, or runs and hides, pretend to ignore him, walk over to the end of the line, stand on it, pick up the line and pull him towards you to recover the toy. (Always remove the line before leaving your puppy alone.)

Remember that puppies' mouths are full of teeth that are either about to fall out or are newly growing. For this reason do not play too roughly. Let your puppy do most of the pulling. You are there to be the anchor and to make the game exciting by

Exercise: Tug-of-war games

▷ **1** Begin by inviting your puppy to pull on the less favoured tug-of-war toy. After a few seconds, bring the toy in towards your body and keep it as still as possible. Holding a treat near his nose may make him let go more quickly. As he begins to loosen his hold on the toy, give your voice-cue (such as 'Drop!').

occasionally pulling gently on the toy. Do not hold the toy up high and wait until your puppy drops off as a means of getting the toy back, because this can cause damage to his mouth.

Your puppy will need to learn to stop pulling and let go when asked, even during a really exciting game. It is helpful to have two tug-of-war toys for the tug-of-war exercise: a favourite one and one less favoured.

At first, only teach this exercise when the game is unexciting – that is, when you have only just begun to play. Follow steps 1 and 2 below.

Gradually start to use the voice-cue during more exciting moments of the game, having first pulled the toy in towards you so that you are able to hold it as still as possible when you give the voice-cue. Work up gradually to saying 'Drop' when your puppy is trying very hard to get the toy. If he does not let go at once, go back to practising at a less exciting point in the game and progress more slowly.

Your puppy will learn that when he hears the voice-cue 'Drop', the game he is currently playing becomes unexciting and that, if he lets go, he receives a treat, praise from you and a more exciting game with a different toy. He will then start to let go immediately whenever he hears 'Drop'.

△ **2** When he lets go, feed the treat, give lots of praise and offer a game with his favourite toy. Practise this often, in short sessions, until he will reliably let go of the toy as soon as he hears the voice-cue 'Drop!'. Remember to reward him well whenever he does so.

'Shake-and-kill' games

Dogs that enjoy killing squeaky toys often have a predatory side to their natures that is easily excited. Since their behaviour with squeaky toys represents what they are capable of, care should be taken when they are in the vicinity of small pets. Teaching some rules to this game will give you the control you need when necessary. Your puppy will need to learn two things:

- To come back when called, even when chasing the squeaky toy
- To stop 'killing' the toy and drop it at once.

The first can be taught using the elementary chase recall technique given earlier (see page 104), but use two squeaky toys instead. The second is easy to teach using the shake-and-kill exercise shown here, provided you begin while the puppy is still young.

Exercise: Shake-and-kill games

▷ **1** Begin by teasing the puppy with the less-favoured toy and allow him to take it in his mouth while maintaining a firm grip on it yourself.

▷ **2** Let him squeak it a few times, then hold a treat in front of your puppy's nose. When he smells the treat it is likely that he will leave the toy to try to take it.

Follow steps 1–3 below to teach your puppy to let go when you ask.

If he is more interested in the toy than the food treat, hold the toy still, but try to squeeze most of the air out of it (so that it no longer squeaks as your puppy bites it) before you try again. Since the toy has now become unexciting, your puppy should become more interested in the treat. If this is unsuccessful, try offering another squeaky toy instead.

△ **3** Give the voice-cue 'Drop!' as he does so. He will eventually begin to associate this voice-cue with taking his mouth off of the toy. Lure him away from the toy slightly and reward him with much praise, the treat and a really exciting game with his favourite squeaky toy.

Practise often, in short sessions, until your puppy is beginning to let go of the toy as soon as he hears 'Drop' in anticipation of getting the treat or his favourite toy. When he is reliably doing this, try again without holding onto the toy. Have him close to you, so that you can hold on to it as before if necessary. Eventually you should be able to do this at a distance. Your puppy will leave the squeaky toy he is playing with when he hears you say 'Drop' because he is anticipating a treat, praise and a more exciting game. Eventually, it will become an automatic reaction.

Other games to play

Once your puppy knows how to play with toys, there is no end to the games that you can teach him. A few are given below, but by being inventive you can modify these games or invent your own. As your puppy becomes 'educated', the games you play can become more complex, which will be more interesting for both of you.

'Find it!'

The aim of this game is for your puppy to find a hidden toy using his sense of smell. It is a really good way to use up your puppy's mental energy before leaving him when you need to go out. It can be played either indoors or outdoors and, once they have learned how to play it, dogs seem to really enjoy it.

Begin by holding your puppy's collar and letting him watch as you throw a toy into long grass or roll it around a chair out of sight. Release him with an excited 'Find it!' and go with him to do so. As soon as he 'finds' the toy, praise him enthusiastically and encourage him to pick it up and carry it back to where you started.

Progress slowly, encouraging him to find toys thrown into many different places out of sight. Then begin to cover his eyes or use your body as a shield when you throw the toy, so that he cannot see where it lands. At first, try to make the toy land with a loud 'thump' when it hits the ground so that your puppy knows it is out there somewhere.

Eventually you will be able to say to your puppy, 'Find it' and he will know that there is a hidden toy somewhere close or underneath something nearby. Always praise the puppy and play with the toy whenever he 'finds' it.

Let your puppy see where the toy is at first before encouraging him to find it when hidden later.

Fetch the slipper

The aim of this game is for your puppy to retrieve a named article. It is another good way to use up mental energy. Although your puppy can't talk, he will nonetheless be able to learn words that are repeated clearly and frequently enough. Since he will be happily playing, he will soon learn the names of the articles you are sending him to fetch and you can develop this game into something really useful, like fetching your slippers or newspaper from the other room.

Begin by throwing out two different articles – for example, a toy and knotted sock. Allow your puppy to run after them and see which one he retrieves. Make a big fuss of him when he returns to you and play a game with whatever he has retrieved. Recover the other object and throw both out again, this time telling him to 'Fetch the ...' as he does so, giving the name of the article he first brought back because that is the one he is most likely to retrieve again. As he brings the correct object back, praise him well and

repeat the name of that particular item again and again. If he brings the wrong object, take it from him without praising and send him for the other. Concentrate on the same object for several sessions until he is reliably bringing back the named object.

You will then need to teach him the name of another object in exactly the same way, using two items different from the first two. Once he knows the names of two objects, throw them both out together and ask him to retrieve each in turn. If he brings the wrong one, do not tell him off, but do not praise him either. Just quietly take the unwanted article, repeat the name of the object you want and send him out for it. When he brings it back, give him lots of fuss and play a game with the object.

Repeated often enough, this game will enable you to teach him the words for as many objects as you have the patience to teach, and it can be a good party trick for him to retrieve the correct item from a big pile of other objects.

Fun games with toys will keep your puppy occupied and out of mischief.

Agility is an activity that is fast, furious and really enjoyed by dogs.

'Grown-up dog' games

Grown-up dogs are often taken by their owners to play competitive games with other dogs and owners. Choose from the following categories.

Dog agility

Dogs learn to negotiate tunnels, hoops, high jumps, long jumps, high walks and A-frames. This is then done around a set course at speed against the clock, as in show jumping for horses. The handler runs round the course and controls the dog's directions, with the dog off lead. Dogs compete against others for the fastest time with the fewest faults.

Dancing with dogs

Dogs are taught 'moves' by their handlers, which are then linked together and choreographed. This can lead to fantastic displays in which the dog appears to 'dance' with the handler. Routines are assessed for 'programme content', 'accuracy and execution' and 'musical interpretation'.

Obedience

Dogs learn to respond precisely to obedience commands, such as 'Heel', 'Sit', 'Stand', 'Down' and 'Stay'. They learn how to retrieve a scented article from unscented ones, and to be sent away to a specified place. Precision is very important and handlers compete for points, which are easily lost if the positions are not exact.

Working trials

This is the (serious) game that police dogs and their handlers often play because it involves many of the elements that make a dog useful to the police. In working trials, dogs learn to track and search for missing objects, are taught how to negotiate a 1.8 metre (6 foot) scale jump and a 2.7 metre (9 foot) long jump, and learn how to walk at heel, retrieve a dumb-bell, come when called, stay when told and be sent away from the handler as directed.

Flyball

This game involves teaching the dog to press a pedal on a box to make a ball fly out of the box, which is then caught and taken to the owner. There are usually a series of jumps set out in front of the box, which the dog negotiates on the way there and back. Once the dog has learned this, teams of dogs race against each other to see which are the fastest.

Gundog work

If you have a gundog breed, you might be interested in exploring gundog work as an activity to get involved in. This does not have to involve shooting or dispatching game if you find that distasteful. Gundog working tests are held in the summer months and comprise a series of identical tests for each dog, based on retrieving canvas dummies and on obedience. These are competitive, with the results being placed.

Water work

If you own a giant breed such as the Newfoundland, you might consider using its natural tendency towards water in a positive way by becoming involved in water work. Regular training events are held in which the dogs can learn to swim and retrieve articles from the water. They can then progress to more difficult tasks, such as towing people and boats, and the advanced dogs learn to jump from a boat and search for a 'casualty' to tow in. These skills can be taught for fun or assessed at progress tests, which the dogs work through at each level.

These games are for grown-up dogs because most involve jumping which is not good for growing limbs. Small obstacles that are very close to the ground can be practised with older puppies, but it is really best to wait until they have matured physically before going

Working trials requires dogs to be able to scale a 1.8 metre (6 foot) obstacle

any further. Other parts of these 'games' can be trained at a very early age, but it should be remembered at all times that these are simply games and are meant to be enjoyed by both participants.

Unfortunately for some dogs, some owners take all the fun out of these games by training too hard. Owners who keep the games light-hearted and enjoyable usually have more success and certainly have a lot more fun than those who must win at all cost. Keep the games fun and you will have a dog that loves to work with you on any task you choose to set.

Flyball is a sport played at high speed. Dogs negotiate small jumps as they race to and from the machine that delivers the ball.

CHAPTER 10
Preventing biting and aggression

This chapter explains the main causes of aggression in dogs and suggests ways to proof your puppy against situations that may cause him to bite. Puppies rarely bite in earnest until they reach the age of about seven months, unless the provocation is extreme or they have a genetic make-up that makes them highly reactive. Before this time, they usually have little confidence in their own abilities and tend to rely on other strategies, such as running away or appeasement.

As puppies mature, and as their confidence grows, they become more likely to resort to aggression to solve their problems. Owners are often unaware that problems are developing until their puppy grows up and becomes aggressive. By understanding why dogs bite, by watching for early signs that all is not well and by removing the necessity for dogs to take matters into their own hands (or mouths!), owners can greatly reduce the chances that their puppy will grow up to be a biter.

Excessive play-biting

The purpose of play-biting is to play, so it is not real aggression. However, puppies have sharp teeth and some have strong jaws, so they can do quite a lot of damage to skin and muscle, especially when they are older, if their 'playing' is uncontrolled.

Puppies will play-bite moving targets naturally when young, and need to be taught to orientate their play onto toys instead or hands, arms and feet (see page 96). This is a relatively simple process, but some puppies will play-bite harder and in a more determined way than others, particularly if this behaviour has been encouraged. It becomes even more urgent to direct the play-biting onto toys if the puppy lives with children or elderly people (who have much thinner skin).

Walk away

To control excessive play-biting, first try the method outlined on page 96. Work hard at this, using a bigger toy if necessary to prevent the puppy biting your skin accidentally. If you are still having difficulty, or if your puppy is determined to bite you, get up and walk away

Standing up and turning away is the best way of bringing an end to the game and your puppy's play-biting.

Uncorrected play-biting will become a problem as the puppy matures. He needs to learn to bite onto toys instead of our fingers.

slowly as soon as you feel your puppy's teeth make contact with your skin. Do not say anything, look at your puppy or touch him. Just walk away so that the game and the social contact are ended. Do this each time he gets over-excited and begins to bite you instead of the toy. Usually puppies learn to stop biting after just a few days of this procedure, since they rapidly learn than biting humans brings an end to the game. Make sure you are playing lots of games with toys, though; otherwise, he will be desperate to play and will continue to try to play in the only way he knows.

Some puppies will be stimulated by the movement of your walking away and may try to bite at your legs or feet. If this happens, stand still and ignore him. If

he is slow to stop, use a house-line attached to his collar to remove him from you, holding him at arm's length until he has calmed down.

'Ouch!'

If getting up and walking away each time he play-bites is unsuccessful after a few days, you may like to try another method. This involves keeping your hand still when your puppy bites, and giving a short, sharp yelp in a loud voice as if you have been hurt (this will be most effective if you do not normally shout at your puppy). Look at your hand and ignore your puppy for a moment. He will probably look startled, wag his tail and attempt to lick your face in appeasement. Praise him quietly and offer a game with a toy. Repeat as necessary, but do not get into a cycle where you shout and then your puppy gets more excited and bites harder. If your first shout does not work, go back to standing up and walking away instead. Remember to praise him and offer a game with a toy whenever he stops biting.

Using a house-line

If your puppy continues to try to play-bite excessively – especially if he is doing it to attract attention, or to prevent him biting young children or elderly people – attach a house-line to his collar whenever you are with him and use it to restrain him. Always have a toy ready so that he can bite onto it, and use the house-line to hold him back so that he cannot bite anyone. At the same time, put the toy within his reach and keeping it moving so that he can bite at it instead.

You can relax a little once he is happily playing with the toy, but use the house-line to move him away again if he begins to bite people instead of the toy, or is getting over-excited.

If excessive play-biting is happening regularly at certain times of the day, try to anticipate the event that sets it off, and restrain your puppy at these times so that he cannot get to you. For example, if he becomes excited when the children come home and tears around nipping them, or begins to bite at you to attract your attention when you sit down to read, put him in his playpen beforehand or make good use of a house-line so that he cannot do this. Ensure he is receiving adequate play and stimulation at other times.

'Mad five minutes'

Puppies are well known for having a 'mad five minutes' every so often, during which they tuck their

Use a house-line for puppies that play-bite hard or excessively, to prevent them reaching your hands.

Puppies are well known for having a 'mad five minutes' from time to time, to use up excess energy and exercise all their muscles.

tail underneath them and tear around like a hound from hell. This is just a puppy way of letting off steam and is perfectly normal. What is not acceptable, however, is if your puppy rushes by family members during these times, biting them on the way past. If this happens, walk out of the room leaving him on his own or, if it is easier, put him out of the room instead. This will help him to learn that such behaviour does not bring him the attention that he hoped for. If he continues to do it, re-examine how many opportunities your puppy is getting for toy play and energy release, both physical and mental, and increase them if necessary. Make sure he is learning to play games with toys gently, without growling or biting, and is not learning to play-fight with humans in any other situation.

Play-biting at clothing, feet and ankles

Most young puppies will be stimulated to play by the sight of a pair of feet walking past them, especially if these are accompanied by a flapping skirt or trouser legs, and particularly if they belong to exciting, playing children. Offered such a temptation, puppies will often run and attempt to wrestle with your ankles, just as they would have done if one of their litter-mates had run past them. Since this is painful and likely to damage clothes, it is unacceptable and needs to be stopped. Provided you are playing enough games with your puppy at other times, you are entitled to make it perfectly clear to him that this behaviour will not be tolerated. Stand still and tell him off loudly. Be scary enough to make him back off as he realizes his mistake. Then praise him for having stopped. Move

Preventing biting and aggression 117

away, and walk past him again later to check that he has learned what you intended. If he makes no move to follow, go back to him and praise him well.

Another way to tackle this problem is to put his playpen in a busy part of the house so that many people walk past it. Since he cannot get out of the pen, he will become accustomed to moving feet and will learn that play only happens when people get toys out.

Why dogs bite

Many people think that dogs bite just for the sake of it, but in fact they do not become seriously aggressive without good reason. Most canine aggression is simply the dog's only way out of a stressful situation. When he finds himself in a predicament that he cannot run from, he cannot sit you down and tell you what he is unhappy about or write a letter. The only means he has of expressing himself is with body signals. Too often, humans ignore these because they do not know how to read them. Should the situation get worse, the dog's only recourse is to use aggression to sort things out.

Many dogs will growl or bare their teeth as a warning before they bite. Unfortunately, owners usually tell the dog off or become aggressive themselves when this happens, instead of realizing that their dog is anxious and doing something to reduce his distress. Once a dog has been taught not to growl or show his teeth in this way, he will often bite without warning, which can be very dangerous.

As well as having a reason to bite, dogs also need self-confidence to use aggression, and this comes with age and experience. Strong-willed types usually have more confidence; naturally submissive types rarely resort to aggression. Dogs have more confidence – and hence are more likely to resort to aggression, if necessary – when they are on their own territory or with other members of their pack (human or animal). Breed differences have a big effect on a dog's threshold for aggression. Some require a lot of provocation to bite, whereas other are more reactive.

Dogs that are protective over possessions are frightening and can even cause serious injury to family members or visitors.

A sideways orientation, squatting down and turning the head and eyes away, allows this shy puppy to come forward to take the treat.

Whether confident or not, dogs do not use aggression lightly and it often upsets them greatly, just as it upsets humans who have to resort to fighting. Dogs do get better at it, however, and will become very skilled at using aggression as a solution if stressful situations present themselves again and again.

The best way to prevent aggression is to proof your puppy against situations that cause dogs to be anxious using the information in this chapter, so that he learns to tolerate or even enjoy them. It is also important to watch for signs of fear throughout your dog's life, to heed warnings (such as body postures, growling or baring of teeth) and to find a way to relieve the anxiety he is feeling during these times.

Aggression towards humans

There are many causes of aggression towards humans, and the secret of prevention is to be aware of all of them and to proof your puppy against normal encounters and take care of him in all situations so that he has no reason to bite.

Fear-induced aggression

Fear is the most common reason for dogs to bite people – both strangers and people who are familiar to them. Dogs often have good reasons for biting, and have usually been pushed way beyond their tolerance level. However, dogs do not need to have had painful or frightening experiences with such people to become afraid. Poorly socialized dogs are likely to be fearful simply because they did not meet enough people and have pleasant experiences with them while they were young.

While a puppy is young and shy, he will usually try to appease, hide behind his owner if he can, or run away from a frightening situation. As he matures, he will become more confident of his own ability to defend himself. In stressful situations from which he has learned there is no escape (for example, when on the lead), he may start to use aggression to try and make the threat go away. Once he finds out how effective this is, the aggressive displays often begin to get worse.

Members of the family who use force and punishment in an attempt to train a puppy are also likely to be viewed with mistrust, as are children who tease or torment puppies, and they are likely to be bitten during confrontations when the dog feels a need to defend himself.

Socialization, protection from unpleasant (painful or frightening) experiences and positive training/education methods are the key to preventing fear-induced aggression. Do not force a puppy into a situation that he is obviously unhappy about. Allow him time to get brave enough to go forward and investigate, and try to make the experience a pleasurable one by using toys and food treats. If he is restricted by his lead or is backed into a corner, help him out by removing the source of his fear so that he does not learn to be aggressive in order to solve the problem. Remember that it is your responsibility, as leader, to protect pack members from harm and help them out of difficult situations, even if this necessitates asking a well-meaning person who is trying to get your puppy's attention not to do so.

Ensure that all unavoidable upsets are countered with many happy experiences in the same situation. If dogs consider all humans – adults and children – to be a source of fun, treats and games, they will not want to 'see them off' with a threat display. See Chapter 6, 'Socialization' (page 54), for ways in which to do this.

Aggression towards children

This is usually fear-induced. Children, particularly toddlers, can appear very different from adults, and sometimes rather frightening from a puppy's point of view. Good socialization with many ages and types of children is, again, the way to prevent problems.

Unfortunately, children often enjoy teasing dogs, and some can be unintentionally cruel. If your puppy has met and had many pleasant encounters with a wide variety of children, he is less likely to feel threatened by the occasional bad experience. A well-socialized, well-adjusted dog will usually be able to deal with a one-off situation in which he is tormented, without resorting to aggression.

Territorial aggression

Territorial aggression is a form of fear-induced aggression. On their own territory, dogs are much more confident about removing a potential threat than they are elsewhere.

If two wolves from neighbouring territories meet, it is likely that the wolf on home ground will win the fight, even though he may be smaller. Being on home ground gives him a psychological advantage because the neighbouring wolf knows he is not supposed to be there. The owner of a territory is also more motivated to defend the resources within it than the intruder is.

However, a dog will attempt to get rid of visitors to a property only if he sees them as a threat. Adequate socialization will ensure that your puppy will be pleased to see visitors rather than want to see them off. This does not mean that he will not raise the alarm. Most dogs will bark when a visitor approaches, and most mature confident dogs will want to defend their property from people who are acting suspiciously, especially if you are not at home at the time.

Unfortunately, delivery people usually seem to be 'acting suspiciously' from a dog's point of view. Not only are they entering your dog's territory, where he is most confident, but there will probably have been many occasions when they have entered your premises, startled your dog into barking, done something 'strange' (such as rattling milk crates or opening the letter box) and then run away as soon as he starts barking.

Although delivery people will move on after they have made their delivery, the dog does not know this. If, one day, someone leaves the door open, the postman may suddenly get bitten on his leg or his backside as he returns down the path. The dog is worried by the postman's daily intrusion and is simply trying to tell him not to call again.

To prevent aggression to delivery people, take your puppy out to meet them often until he is mature (even if it means getting up very early in the morning!). Encourage delivery people to offer food treats and to throw a toy, if they have time. Usually they are only too pleased to make friends with a puppy, especially if you explain why.

Territorial behaviour, such as barking at the gate, usually begins as the puppy gains confidence and begins to mature after puberty.

If necessary, a weatherproof box by your gate can be used to store food and toys, and delivery people can often be encouraged to give these to your puppy occasionally as they make their deliveries. A few treats put through the letter box with letters could encourage a puppy to be very pleased to see the postman at the door, and to begin to look forward to their arrival rather than being worried enough to try and scare them away. Most delivery people are happy to do this, rather than have to face an aggressive dog every morning (even if it's on the other side of the door). This will not prevent your puppy from alerting you when someone comes to the door when he is older, but it will stop him from growing up to be worried about delivery people and aggressive towards them.

Aggression in the car

Small territories, and ones with very definite boundaries, allow easier defence from threats. This is one of the reasons why some dogs become aggressive in cars. This problem also has its origins in fear. The dog is not so much protecting the car as protecting himself. He has no means of escape and therefore needs to threaten intruders to keep them away and, since the car is a small and easily defended space, he feels confident about doing so.

Again prevention lies in good socialization. If your puppy has no fear of people, he will not feel threatened when they approach. For shy puppies, it may well be worth asking people to go up to your puppy when he is in the car to offer food treats and games, so that he begins to see them as a source of

Exercise: Preventing food aggression

▷ **1** Call your puppy to you and place his dinner in a large bowl on the floor. Squat down beside him and, as he begins to eat, gently stroke and touch him all over, talking to him quietly. Begin on his back and side and, if he is comfortable with this, progress slowly towards his head. Stroke his ears and underneath his neck if he is relaxed. Take small pieces of food that are more appetizing than his ordinary food (for example, little bits of chicken, liver and fat trimmings) and hold them in front of your puppy's nose as he is eating. Allow him to eat these tasty food treats. Do this several times.

pleasure rather than viewing them with mistrust, which may lead to aggression later on.

Food and possession aggression

Preventing food- and possession-guarding is quick and easy to do if you start early enough. It is easier with puppies that were fed individually while still in the litter and given enough things to investigate and play with; puppies that had to fight over a communal bowl to get enough to eat, or over possessions, will have learned to guard early in life and are a little more difficult to deal with.

Food aggression occurs because of a need to protect a vital resource. It is a natural and normal behaviour, but one which can (and should) be eliminated in a pet dog. It is not a problem that is

Take care!

- If your puppy is already aggressive, ask for expert help from an experienced behaviourist so that you do not risk getting bitten (see page 203).

related to status; a subordinate will often guard an item of food from a higher-ranking, stronger animal. A law of possession seems to apply. Dogs that have been hungry at some point in their lives are more prone to food-guarding than ones that have always had enough to eat because they have learned to protect the food they acquire.

△ **2** Prepare another piece, but this time hide it in the palm of your hand. Put this hand, palm up, in the centre of his dish and, as he begins to sniff at it, open up your fingers to allow him to find and eat the food treat. Do this several times, sometimes digging around in his dinner bowl before offering the food treats, then leave him alone to finish the rest of his dinner in peace.

△ **3** When your puppy wags his tail as you approach, occasionally lift his bowl up, put a few tasty food treats in it and return it at once. Once he is happy to let family members approach his food in this way, get other people to do it while your puppy is still small, especially children (particularly if there are none in your family). Try to do this every day until your puppy begins to wag his tail and look up from his bowl to see what else is on offer as you approach, then reduce it to just a few times a week until he is mature.

To prevent food aggression, you need to teach your puppy that humans approaching are not coming to take the food, but are instead bringing something more appetizing to eat (see previous page). Once a puppy has learned this, he will welcome you rather than trying to keep you away.

Never become aggressive yourself in response to food aggression, or you will teach your puppy to become much more aggressive next time in order to protect his food. Eventually this will cause a rapid escalation in the aggression until your puppy becomes dangerous whenever he is eating.

For puppies that are not already food-possessive, follow the preventing food aggression exercise shown on page 122 at meal times.

Bones and chews

The same procedure is required for bones and chews, using the exercise shown below. Always give the bone or chew back immediately, and he will learn that you can be trusted and there is no need to guard it from you. Ensure that everyone in the family can approach your puppy while he is chewing a bone or chew without provoking aggression. If you have

Exercise: Bones and chews

▷ **1** Give your puppy a sterilized marrow bone or chew and leave him until he is just beginning to get bored with it. Approach him calmly and confidently, holding out a tasty, smelly piece of food that dangles from your fingers.

▷ **2** Give him time to smell how nice it is, then lure him away with this food and pick up the bone or chew as he moves away from it.

children, supervise these sessions carefully until your puppy is happy to let them approach too. Always ensure that they have very interesting, tasty food treats to give, and do not allow them to continually pester your puppy with this treatment. Teach them to respect other dogs with bones and not go near them because not all dogs will have had the benefit of this experience.

Toys and other objects

In a similar way, you can teach your puppy not to be annoyed about you taking objects that he is playing with. At first, always offer something better in exchange when you take something from him, such as a tasty treat or a game with a better toy. As he matures, reduce the rewards you offer, but try to keep him guessing about whether or not you will give him something nice in return. By doing this, you will teach him to be happy for you to take something from him and he will not try to avoid you, be aggressive or run away so that you cannot take it. Teaching him to retrieve (see page 98) will also help with this process as it allows him to practise and get used to bringing things to you.

▷ **3** Allow him to eat the treat, then immediately return the bone or chew to him. After a few repeats, he will begin to look up at you expectantly when you approach. You can then begin to offer him the treat with one hand while taking his bone or chew with the other. Eventually you will be able to take his bone first and then open your other hand to reveal the treat.

Pain-induced aggression towards humans

Dogs will often bite if we approach them when they are in pain. In trying to help them, we may have to cause them more pain temporarily and they often bite in an attempt to make us stop. Little can be done to prevent this, apart from trying to protect your puppy from illness and injury – as all good owners will seek to do anyway. Routine handling and grooming exercises will help to build up trust between you so that, if accidents do occur, your puppy is more likely to let you help (see page 141 for advice on how to get your puppy used to wearing a muzzle, in case he has to wear one in an emergency).

Dominance aggression towards humans

Dominance aggression is usually directed towards members of the family or people with whom the dog spends a lot of time. Only ambitious dogs are interested in taking control, and before they do so they will have assessed the strength of the people they are challenging. For this reason, this problem is rare. The prevention of dominance aggression lies in putting into practice all the procedures outlined in Chapter 8 (see page 78). Puppies brought up to respect the humans in the family will not challenge for leadership.

Chase aggression

Chase aggression occurs when a dog finds an outlet for its desire to chase by running after unsuitable moving objects. The 'prey' may be cats, sheep, ducks, deer, horses, rabbits or even small dogs. In the absence of 'prey' animals, and without adequate opportunities to reorientate onto a toy their desire to chase, dogs may often run after other fast-moving objects, such as joggers, cars or people on bicycles.

Dogs that are particularly prone to this are those from the herding breeds or hounds, which are very stimulated by movement.

Usually the chase is enough in itself and the dog will pull up if the creature stops running. However, not many animals can be chased without becoming frightened, and this usually means that they try hard to evade capture, making the chase even more exciting. If an inexperienced dog catches up with a fleeing creature, he will usually pounce on it to stop it running. Once caught, frightened animals tend to become aggressive, and this only needs to happen a few times for the dog to learn to bite first as he catches up with his quarry, to save himself from being attacked.

Dogs with little to do and not enough control will rapidly learn to nip the legs of joggers, attack little dogs in the park, worry livestock and attack cats in other people's gardens. Prevention involves socialization, control and reorientating onto toys your puppy's desire to chase. Never give a puppy unsupervised access to prey animals by allowing him to stray or wander far from you when out on a walk. Develop the chase recall (see page 104) so that, should your puppy unexpectedly set off after something, you have the ability to call him back.

Remember that dogs are often closer to their wild ancestors than we like to think, and should never be left alone with small pets such as hamsters, guinea-pigs or rabbits.

Aggression towards other dogs

Understanding the causes of aggression to other dogs will help you prevent your puppy from developing problems with other dogs in later life.

Fear-induced aggression

Aggression towards other dogs is often caused by fear. Inadequate socialization or bad experiences can cause a puppy to grow up being very fearful of encounters with other dogs.

As a fearful puppy matures, he will begin to use aggression to keep other dogs away, particularly

Chase games between puppies are usually harmless but keep an eye on proceedings to make sure both dogs are enjoying the game.

when he is on the lead because he will have learned that, once on the lead, there is no way to escape. Such aggression is usually made very much worse by owners tightening the lead and becoming anxious themselves whenever they see another dog coming towards them. Off the lead, such puppies will often keep their distance from other dogs, although a few want to be friendly. They will run up to other dogs, only to find that they are out of their depth, and then they resort to aggression to resolve the situation. To make matters worse, puppies that have been poorly socialized often unwittingly send out the wrong signals because they have not learned to use their body language correctly. This causes some dogs to become aggressive towards them, which simply aggravates the situation.

Socialization and protection from bad experiences are the key to preventing fear-induced aggression towards other dogs. If your puppy is attacked or frightened by another dog, ensure that he receives many enjoyable encounters with similar dogs in similar situations to overcome the unpleasant experience and to erase the mental scar.

Rough play

Some dogs will fight with others because they have learned to play roughly as puppies. This applies particularly to puppies brought up with an older, tolerant dog, which has allowed them to play rough games without correction.

Puppies that learn to play very rough games will continue to do so as adults. Since few dogs will tolerate such behaviour from a dog they do not know, rough players will often meet with aggression from the other dog. Once this has happened a few times, the rough player learns to weigh up the situation and to launch a defensive attack if the other dog looks as though it will be aggressive to him.

To prevent your puppy turning into a dog that enjoys rough games, stop him playing as soon as his games with other dogs begin to get out of hand. Do not let him do anything to a familiar dog that he would not get away with doing to an unfamiliar dog. This includes biting them hard, putting his paws on their back and mounting them. Interrupt your puppy if he does this, get his attention for a while and let him calm down, before allowing him to go back and play.

Play sessions with other pups should be short, friendly and not too rough.

Aggression between dogs on leads is common and often due to fear caused by lack of early socialization. Help your puppy by gradually exposing him to other animals.

Stop the session completely if he plays too roughly or gets over-excited.

Always ensure that your puppy would rather play with you than with other dogs by reducing the amount of time he plays with them if necessary. Practise with a friend's dogs, asking them to stop their dog from playing while you get your puppy back. Reward him well before he goes back to play.

Inter-male aggression

When male puppies become sexually mature, some entire (un-neutered) male dogs may become aggressive towards them. This should not be too much of a problem if your puppy is well socialized, since he will know the body signals necessary to appease the other male. However, it will be more of a problem if you live in an area where several aggressive entire males use the same small exercise areas.

Try not to worry too much about the odd aggressive incident; if you are anxious, your puppy will sense this and become anxious, too. Instead, concentrate on playing with him when the other dog comes into view, so that his attention is fixed on you. Keep him on the lead, and keep your distance from the other dog, turning round and walking away from the situation if necessary. Try to avoid aggressive entire males during the adolescent phase, and keep up your puppy's socializing with friendly females and neutered males.

If your puppy is male and poorly socialized, you may find that he becomes more aggressive to other entire males as he matures. If the aggression continues into adulthood, you will need to find an experienced pet-behaviour counsellor (see page 203) to help you overcome this, and you may also need to consider castration.

CHAPTER 11
Chewing

Puppies chew because they are teething, to exercise their developing jaws, for fun and to find out about their environment.

Teething occurs between the ages of three and six months. During this time most puppies will have an uncontrollable urge to chew to relieve some of the discomfort and to facilitate the removal of puppy teeth and eruption of the adult set. Just as the teething phase begins to pass, another more ferocious urge to chew occurs. This stage starts at about seven to ten months of age and can last for

about six months. It tends to be more pronounced in the gundog breeds, particularly Labradors, perhaps because they were bred for their ability to use their mouths to carry things.

This later chewing phase coincides with a time when, in the wild, puppies would be starting to travel away from the den area to explore their new environment. Most puppies in pet homes find themselves confined to a relatively small territory, so the desire to explore gets channelled into chewing objects instead. Since puppies have no sense of

Sturdy rubber toys filled with interesting treats are safe to chew and will keep a puppy occupied for hours.

Providing chews suitable for teething puppies, such as rawhide, sterilized bone and kongs, will help prevent damage to household items.

value for objects, and they are now bigger with stronger jaws than when they were teething, they are capable of doing a great deal of damage in a very short time until this period has passed.

Teaching right from wrong

You will need a supply of items that are safe and tough enough to survive being chewed. These can be rawhide chews, smoked bones, Nylabone chews, sterilized marrow bones and strong toys made of hard rubber. Keep them interesting by stuffing food, cream cheese, peanut butter or other tasty food treats into the toys or bones.

Be inventive and make sure you have at least two or three items for each day of the week. This can be expensive initially, but is cheaper in the long run than replacing chewed household items. You can then give your puppy a variety of chew toys each day, putting them away at the end of the day and giving different ones the next day. In this way it will be a week before your puppy sees the first chew toys again, so he will be very interested in them. Throughout your puppy's first year try to provide a variety of things to chew. Ring the changes and keep them as interesting as possible.

In addition, you can give him things to use up his desire to explore. These can be items like cardboard

Hide toys, food and interesting objects in an old cardboard box to keep your puppy busy when he's left alone.

boxes, cereal packets, plastic milk bottles, old plastic crates and wooden logs, which puppies will clamber over, chew on and investigate. Encourage him to find tasty treats hidden in these to maintain his interest. Replace and change these items frequently. Supervise your puppy while he has them and remove any that become damaged and may be likely to cause harm.

In order to minimize the damage done by your puppy during these stages, it is important to deny him access to places where there are valuable or dangerous objects whenever you are not there to supervise him. The easiest, and ultimately cheapest, solution is to invest in a playpen where he can go when you are not watching him (see page 45). This can be filled with a variety of objects for him to explore and exercise his jaws on.

Supervise your puppy's excursions into the rest of the house so that you can teach him the difference between the right and wrong things to chew and instil the correct habit right from the start. Do this by picking up any small household items he may be attracted to, and by providing your puppy with one or two chew toys that he has not seen for a while. Leave

them out on the floor whenever your puppy is in the room. When you see him settle down to chew one, get down beside him and praise him well. Then allow him to chew without interruption.

If he ignores the chew toys and wanders away to chew something else, you need to interrupt him and encourage him back to his chew toys. If this happens often, either replace the chew toys with something more interesting or give him more activity in the form of play, exercise and the chance to explore. In this way your puppy will quickly get into the habit of chewing only chew toys and will begin to leave other items alone.

Until he is reliable, never leave him alone with the run of the house. If he chews something he should not, consider it your fault, not his, for trusting him too soon. There are bound to be accidents, just as there are in housetraining, when you have a lapse in concentration. Do not shout or punish him. This will simply cause him to mistrust you. Just be more careful next time.

How long this process takes depends on how active your puppy is and how conscientious you are.

Some puppies have a much stronger desire to chew than others and will take longer to learn. However, no puppy should really be trusted completely until he is at least one year old and has gone through both stages of chewing. Until that time, always provide plenty of appropriate chews, taking them away and providing different ones at the end of each day to keep them interesting.

When your puppy becomes an adult, his desire to chew will diminish, but it is important to continue to give him bones and chews throughout his life to exercise his jaws and keep his teeth clean.

As well as providing plenty of things to chew, the chewing phases will pass more quickly if you give your puppy lots of mental and physical exercise, as well as the opportunity to explore different environments. Bored puppies kept in one place tend to use up their desire to be active by chewing more.

Learning to be left in the car

Even a small amount of chewing damage done by a dog left in a car can be very expensive. Be careful during the chewing stages and always leave a new and interesting chew toy for your puppy to concentrate on whenever you leave him. Alternatively, get him used to being in a travel crate and leave him in that with a few toys and chews. Accustom him gradually to being left in the car so that he does not become anxious (see page 152). Never leave a puppy or adult dog in a car on a sunny or warm day.

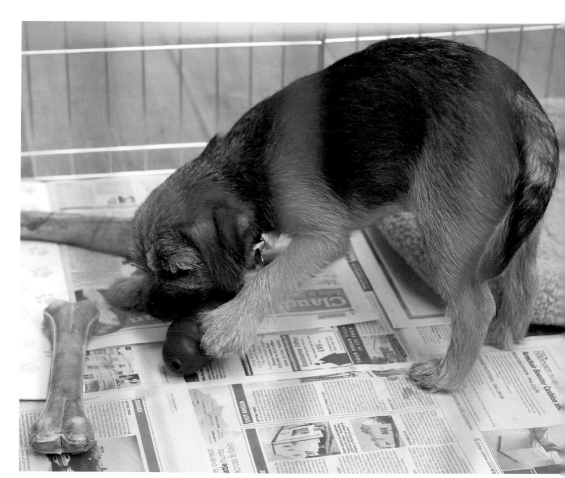

Access to lots of chew toys in the playpen will make a puppy less likely to chew when he's in the rest of the house.

CHAPTER 12
Handling and grooming

Humans are primates and we like to touch, hold and hug. Dogs rarely do this to each other and need to be taught to tolerate and enjoy it when humans do it to them. Even if you and your family do not intend to touch your puppy much, you still need to proof him against the occasion when children come running up and throw their arms around him or when an adult, who should know better, pats him heavily on the head.

The more puppies are familiarized with being held, handled and restrained, the less they feel threatened by such experiences and the less likely they are to bite when touched, particularly during stressful situations such as during an uncomfortable procedure at the vet's clinic. Touching, holding, restraining and hugging your puppy builds up acceptance and trust that will be projected onto people outside the immediate family.

Children like to hug so prepare your puppy for this when he is young by familiarizing him with being handled.

Teaching your puppy to accept gentle restraint is a valuable lesson for a time when he needs to be examined.

Gentle restraint

Before you can handle your puppy successfully, he will need to learn to accept being gently held when he wants to be free. Practise in short sessions, holding your puppy gently, but securely. The more secure he feels, the less he will struggle. Try not to squeeze him or dig your fingers in. Hold him so that he cannot bite at you, and wait for him to calm down. Hold on tight if he struggles and wriggles, but as soon as you feel him relax, let him go. He will quickly learn that struggling gets him nowhere, but keeping still wins him freedom, and you will gradually be able to hold him for longer periods of time.

Everyday handling

Take every opportunity to handle and touch your puppy all over. Sit down with him and keep him calm until he has relaxed. Then begin by touching him gently on the head and back. Once he starts to accept this, progress to more sensitive areas, such as around his face, paws and under his tummy. Progress gradually, keeping your hands slow, and move on to new areas at a speed with which he is comfortable. Give him a soft toy to chew on to prevent him biting on you. If he leaves the toy to bite at your hands or tries to pull away, you are probably going too fast. Slow down and be more gentle until your puppy relaxes again.

After a few sessions, when he is comfortable with being touched and stroked all over, begin to do things that will allow for easy veterinary examinations later. Look in each ear and gently lift his lips to look at his teeth. Turn him round and lift his tail gently. Lift him onto a table and repeat these exercises. It only takes a few minutes to do this and, if you do it every day, your puppy will learn to accept it completely and even enjoy it as it will be a time when he has your undivided attention. Open his mouth by gently prising his front teeth apart and let go at once. Gradually work on this a little every day, until you can hold his mouth open to inspect his teeth without him worrying or struggling.

Most dogs need to have their nails cut periodically. How often depends on the shape of their feet and how much walking they do on hard surfaces, but it is as well to familiarize puppies with the procedure. Ask your veterinary surgeon to show you how to clip your puppy's nails when you take him for his first vaccination; get him to show you the procedure even if they do not need doing at that time, so that you can accustom your puppy to it while he is still young. You can practise this daily without actually cutting them, so that he gets used to this as part of his routine.

Although handling sessions should be enjoyable, they should not have the atmosphere of a game. Your puppy needs to be taught to be sensible during them, and although care should be taken never to allow him to become fearful, neither should these sessions

Exercise: Handling

Handling exercises will ensure that your dog feels less fear and will be easier to handle when your vet needs to examine him.

▷ **Lifting** Get your puppy used to being lifted up onto a table. Support his weight with one hand under his bottom and the other around his chest. Hold him firmly, but not tightly, and keep him close to your body rather than letting him dangle in mid-air.

◁ **Paws** Stroke gently down each leg, slowly pick up each paw, and separate the toes. Talk to him quietly while you do this. Do not be bossy if he tries to pull away (puppies will instinctively pull their paws away when they feel them being held). Repeat whatever it was he did not like, but do it more slowly and gently until he accepts it.

be too much fun, since you do not want your puppy to become excited. Happy acceptance of the situation is what you should be aiming for.

All of the handling exercises are useful preparation for the day when something is wrong with your puppy. If he has almost eaten a fish hook and it is caught somewhere in his mouth, or he has a piece of stick lodged at the back of his throat, he is much more likely to allow you or the vet to extract it if he is familiar with having his mouth opened. If he has hurt himself, he will be much more willing to be examined if he trusts people and has no fear of being handled at other times.

When lifting or manoeuvring your puppy during handling exercises, ensure that your hands are kept flat. If they aren't, your fingers will press into your puppy, causing discomfort, and he will try to wriggle away from you. Flat hands mean that the pressure is evenly distributed, making the exercise more comfortable for your puppy.

▽ **Eyes** Gently wipe around your puppy's eyes as if to remove any matter from the corners.

△ **Ears** Get him used to having his ear flaps held up while you look inside. Occasionally wipe the insides gently with moistened cotton wool. (Never put anything in the ear or poke about inside it.)

▷ **Mouth** Lift his lips to look at the sides of his teeth. Open his mouth occasionally to look inside. Do this by gently prising his jaws apart. Open it just a little way and only briefly at first, and praise him afterwards. As he becomes more familiar with this, it will become easier to do and you will be able to open his mouth for longer periods.

Grooming

This is a more formal way to teach your puppy to accept being handled and restrained. Begin grooming as soon as you get your puppy, and be sure to do this every day whatever the length of your puppy's coat. Use up some of his excess energy first with a short play session. Start grooming your puppy in very short stages – just enough to touch your puppy once all over – and progress gradually to longer sessions.

If he tries to bite at the brush, tuck a finger in his collar and gently turn his head back to face the front. If he tries to sit down when you want to brush underneath him, gently lift him up into the standing position again. If your puppy is very active and you find it difficult to brush him, slow down. Slow hands result in a calm puppy, so try not to brush him too fast, and be very careful not to pull hard when removing tangles in his coat.

When your puppy has stood still for a while and accepted what you are doing, break off suddenly and reward him with his dinner, a game or a walk. Once your puppy understands that standing still brings freedom, he will be more likely to do so and will learn to relax until you have finished with him.

Exercise: Grooming

▷ Hold him gently but firmly, with a hand round the front of his chest and shoulders. With a soft brush, begin along his back, down both sides and then between the back hind legs and underneath his body. Brush each leg and then his head. Be gentle near the head area as it is very sensitive. If your puppy wriggles and squirms at any time, hold on firmly, using both hands if necessary, until he stops. Talk to him all the time in a quiet voice, praising him whenever he stands still.

Occasionally put your puppy on a table to groom him. Follow the advice given on page 136 for lifting him. This will allow him to get used to the experience, which will be beneficial during visits to the vet or the groomer. Make sure he doesn't fall off the table.

Teach him to accept being dried with a towel by gently rubbing him down. Go more slowly if he starts to get wriggly, and keep sessions very short at first, building up the length of time over which you do this.

If your puppy has a long coat that will need to be clipped, get him used to the clippers early on by holding against him either the clippers or an electric shaver without blades while you feed him something nice. Get him used to hair dryers in a similar way, but try to avoid blowing his face until he is very used to them.

Looming and grabbing

Although most owners know how to approach dogs correctly, some people will unwittingly try to touch a puppy in a frightening manner. They will, for example, loom over a puppy, staring straight at him, and then bring down a hand on top of his head. Getting a puppy used to this will often happen naturally as part

Exercise: Grabbing

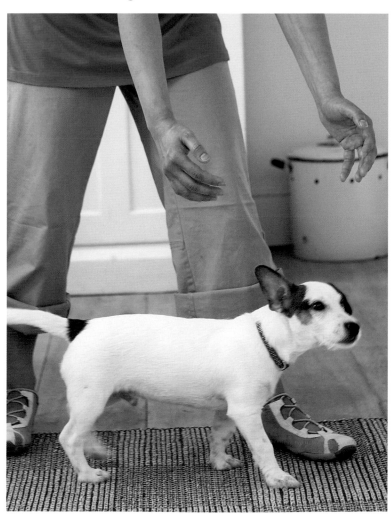

◁ Help your puppy to become used to people grabbing him by making the exercise fun. He will then learn that this is normal human behaviour and will not be afraid. Start by grabbing lightly and gently around the neck area and gradually work up to grabbing him more vigorously.

of the socialization process, but it is as well to proof your puppy against it, just in case.

Try to make this exercise fun and part of everyday life. Occasionally loom over your puppy, extending a hand to pat him on the head. Then offer a food treat and have a game with him. He will soon learn that this is normal human behaviour and will cease to be worried by it.

In emergencies, people often need to grab at dogs, usually around the neck area, either for their own safety or that of others. If unfamiliar with this

Avoid the head!

• Puppies will not appreciate physical affection from you if it means that their head area is touched, as this covers up their important sense organs and touches their sensitive whiskers. Instead, stroke the chest area gently or the neck and back areas.

Stoking the chest area allows this puppy to feel more comfortable than if hands were placed on his head.

Grooming will be a more pleasurable experience for your puppy if it is gentle and slow.

procedure, a dog may turn in self-defence and nip the hand that does it. While your puppy is still small, proof him against this by gently grabbing him and offering him a food treat as he turns to see what is happening. Gradually work up to grabbing him more vigorously – but never so that you hurt him. Always have a food treat ready and make it into a game by praising and playing with him afterwards. The neck is particularly sensitive to being grabbed as this is a place that other dogs will often attack. Since it is also the place where most people will grab a dog, pay particular attention to desensitizing this area.

Handling by other people

Once your puppy is familiar with the handling and grooming procedures, ask other people to try them. Start with people whom your puppy knows well and is friendly with, and work up to total strangers. Supervise this constantly and help out if your puppy begins to look anxious. The generous use of food treats during these exercises will ensure that your

puppy learns not just to tolerate them, but also to enjoy them. It is also important that he gets used to being handled by children so try to involve them, too.

Muzzles

A well-balanced, well-adjusted dog should have no need to bite. However, even the sweetest-natured dog may do so if in severe pain. For this reason it is a good idea to accustom your dog to wearing a muzzle from an early age. Open basket-type muzzles are best for dogs as they can pant and drink with them on. Introduce one by feeding treats from it at first, and by not doing it up until later when your puppy has got used to it. Then put it on briefly, just before something nice happens, such as a family member coming home, going for a walk or dinner. Take the muzzle off as soon as the good thing begins to happen. In this way it will be linked to the arrival of pleasant experiences and will be tolerated more readily. Buy two well-fitting muzzles when your dog is an adult, and keep one in the car and the other at home in case of emergencies.

Handling and grooming **141**

CHAPTER 13
Good manners

It is much easier to create good habits than it is to cure bad ones. Good habits that are developed while your puppy is still young and impressionable will last a lifetime. Establish them by manipulating situations so that he finds it more rewarding to do the right thing than to do something you don't want. You can then reward good behaviour, which increases the chances that it will occur again.

Bad habits are equally persistent so it is as well to prevent them from happening. If your puppy is completely stopped from doing things that are unacceptable for his first year, he will not do them when he is older. Behaviour you do not want, like jumping up, is often rewarding so anticipate it and take steps to prevent it, to ensure that you never give your puppy the opportunity to learn it is worthwhile. This will give good habits a chance to become established.

It is wise to decide now, as a family, what behaviour is acceptable, to foster this and prevent your puppy doing things that you will not like him doing later on.

Learning to deal with frustration

Pet dogs cannot be free to do just what they like in a human world, since their freedom is restricted by fences, doors and the wishes of their owner. It is best that your puppy learns to deal with the feelings generated by not getting his own way sooner rather than later. Just like angry toddlers wanting sweets they cannot have in supermarkets, your puppy may show unwanted behaviour when prevented from getting something he wants. It is much better that he learns self-control gradually and on home territory rather than in a public place where it will be hard to ignore. It is also important that he learns to deal with frustration early on, before he gets larger and becomes more difficult to control.

Teaching puppies to deal with frustration is really easy and should be a gradual process that happens naturally as the puppy develops. However, if you have a strong-willed puppy and you are a very indulgent owner, you may like to try the following exercise to ensure your puppy has learned the lesson.

Teaching a puppy to deal with the frustration he feels when he cannot get his own way is an important part of his education.

A puppy who has been taught good manners will be more welcome than one that continually jumps up.

Find something that your puppy wants and, with him attached to you by a collar or harness and lead, throw or place it just out of his reach. Ignore anything he does, such as struggling, wriggling, biting at the lead and vocalizing. Don't say anything, touch him or tell him off, and don't let him move forward. Wait until all the excitement has died down, he begins to calm and his behaviour becomes acceptable, then let him free to get what he wanted.

Practise this daily, working up from things he is not too interested in to things he really wants. Once he remains calm at home, practise it in other places. In addition, take all opportunities to do this exercise when excitement levels are high – for example, if other puppies are playing nearby and he wants to join in, wait for him to be completely calm before you let him free. In this way he will learn to deal with his feelings when he cannot get his own way and will learn to control himself in all circumstances. This will result in a calm puppy that waits patiently when you want him to. Such self-control is the basis of all good manners.

Jumping up

It is natural for a puppy to want to get closer to your face when greeting you, and jumping up is his way of doing this. If you talk to him, look at him or reward him by giving him any attention when he does this, even if you tell him off, he will learn that it is effective and it will become an annoying habit.

To prevent this, *never* greet a puppy unless all four of his feet are on the floor. If he is jumping up already, do not pay him any attention, but instead turn away slowly and keep still until he drops to the ground before greeting him. Then crouch down and put your face down to him so that he can greet you properly (he may lick your face, so be prepared!). If he is small, you will need to get right down to his level. This will prevent him from jumping up, and if jumping is *never* rewarded, he will quickly learn not to do it.

If you cannot greet him straight away – for example, if you have just entered the house and are carrying shopping bags – ignore him completely until it is convenient. Walk quickly past him, being careful not to step on him, until you are ready.

Since puppies are cute and cuddly, other people will want to greet your puppy, too. They will quickly undo all your good work if they allow (or even encourage) jumping up. For this reason, always supervise your puppy when meeting new people, and either prevent him from jumping up by holding his collar or teach the person how to greet puppies properly. Pay particular attention to the way your puppy greets children. Children usually cause great excitement in a puppy and they are less likely to know what to do so they can teach a puppy to jump up very quickly. Supervise things so that your puppy learns to treat them with respect.

Exercise: Greeting people

▷ **1** Begin with your puppy on the lead when you greet people so that you can control what he does. Your puppy may attempt to jump up at the visitor but, since you are holding him on the lead, he will be unable to do so. Your visitor should wait until all four of the puppy's feet are on the ground, and any excitement has died down, before greeting him.

Children in your family need to know what to do if the puppy jumps up during play. Teach them to fold their arms, stand very still and look straight ahead rather than at the puppy, while saying 'Off!' loudly. Since they are now still, the puppy will lose interest in them and will wander away to find other entertainment. He will eventually begin to associate the 'Off!' voice-cue with the end of his game and will in time learn to respond to the word itself.

Greeting people

The exercise shown below will teach your puppy how to greet people politely. After a few repetitions, your puppy will begin to calm automatically when greeting people, although you will have to supervise meetings throughout his puppyhood just in case. Practise this

particularly at the front door, so that your puppy learns how to behave well when visitors arrive. Arrange several sessions where all the family go out through the back door and in through the front door one at a time, ringing the doorbell and being invited in as if they were visitors. The neighbours may think you're mad, but you will eventually have a dog that waits to be greeted rather than one that jumps up or behaves badly.

Once your puppy has learned to sit on request, you can teach him, in a similar way, to sit before greeting visitors which is even more polite.

Walking past people and dogs

All the socialization you are giving your puppy will make him very pleased to see new people and dogs.

▷ **2** As soon as your puppy is waiting calmly, the visitor should immediately come forward, crouch down and reward the puppy with praise, a stroke under the neck and a tasty treat. Hold on to his collar at this stage if necessary so that you can prevent him jumping up if he attempts to do so.

If your puppy gets excited again and tries to jump, the visitor should withdraw their hands and ignore him completely. As soon as he has calmed down, the visitor can then give him attention again.

Teach your puppy to come back to you *before* he is allowed to greet a stranger or another dog.

However, there will be some you do not want him to greet, such as people who are afraid or simply do not like dogs, or a dog that may be aggressive.

You should, therefore, teach your puppy to come back to you first before running to greet someone or another dog. Call him to you whenever you see a person or dog approaching and reward him well when he gets to you. You can then let him play with suitable people or dogs and keep him away from others. Eventually he will learn the routine and will begin to return to you whenever he sees someone or a dog approaching. If you decide to keep him away from whoever is approaching, encourage him to concentrate on you by playing with a toy or by offering a food treat until you are past them. By doing this he will learn to walk calmly beside you past other people and dogs, rather than straining and pulling to get to them because they are more interesting.

Excessive barking

As puppies mature, they begin guarding naturally at the age of about six to seven months. Owners who are unaware of this, and who encourage their puppy to bark at a very early age, end up with a dog that barks at the slightest opportunity. This soon becomes a nuisance and is difficult to control later.

Never encourage a puppy to bark. If you have done enough socialization, your puppy will be relaxed and happy and will not regard familiar noises and visitors as a threat. This does not mean he will ignore a serious threat to you or your home. Well-socialized adult dogs seem to know when to protect the pack or territory and will do so without being taught.

Try not to let your puppy bark as he runs into the garden, as this can be annoying for neighbours and may become a bad habit. If he is excited about going to the garden and is likely to bark, put him on a lead and do not move forward until he is quiet. Accompany him when he is in the garden if necessary, and interrupt him if he barks unnecessarily. In this way you will teach him to be quiet in the garden unless there is something serious to bark at. You will eventually be able to let him out on his own, but only once he has learned acceptable habits.

Excessive barking is a nuisance and hard to control in adult dogs so it is best not to encourage the behaviour in puppies.

It is natural for some dogs, particularly the smaller breeds, to bark when excited. Never reward this by giving your attention, talking to your puppy, shouting or touching him. If you encourage this when he is a puppy, it will get worse as he gets older. Instead, look at the situation that prompted the barking and try to work out a way to reward him only when he is quiet. Say, for example, that your puppy barks with excitement when he thinks he is about to go out for a walk. Make sure you do not reward this behaviour by continuing with the preparations for a walk while he is still making a noise. Instead, stand still, ignore him and wait patiently, avoiding eye contact. Only proceed with the walk when he is quiet.

Taking food gently

Puppies will quickly learn to snatch at food you are holding if you pull your hand away as they approach. For this reason, keep your hand still when you are offering food so that your puppy can learn to judge where to aim his teeth. At first he will be clumsy and will probably try to chew your fingers as well, although he will not mean to. Try not to move your hand and he will soon learn to be more gentle. If you cannot tolerate this, put food treats on the flat of your hand at first and gradually progress to holding them in your fingers.

Another way to teach this is to hold the food treat between your thumb and finger, but curl your hand

Exercise: Taking food gently

▷ **1** Hold the food treat between your thumb and finger, but curl your hand round to make a fist before offering him the back of your hand. As he comes forward to snatch the food, he will bump against the back of your hand.

round to make a fist that the puppy bumps into, as shown in the exercise illustrated below.

Children should be taught these methods of giving food treats since their fingers are often more sensitive and they are more likely to pull their hands away, encouraging your puppy to snatch more hurriedly next time before the treat is taken away. Try to prevent situations where he will be able to snatch food from a young child who is wandering around eating. Either sit the children down or restrain your puppy until the food is eaten. Never tease a puppy with food as he will learn to lunge for it, which could result in an unintentional bite. Always give food confidently so that they do not learn to snatch.

No begging

If you feed your puppy while you are eating or just afterwards, he will quickly learn to anticipate being fed at this time and will sit beside you drooling and dribbling. Preventing this behaviour is easy if you make it a rule of the house that no one is allowed to feed him at these times. Your puppy will then learn that he cannot have human food and will ignore humans when they eat.

Puppies also need to be kept away from the table while you are eating so that they are not rewarded by bits of dropped food, or by those given on purpose by the children or visitors who may be tempted to do so. Teach your puppy to lie down at the side of table and

◁**2** He will hesitate and, as he does so, uncurl your hand to present the food on your flat fingers. He will then take the food treat without biting your fingers. Eventually he will learn to come forward and wait close to your hand for you to open it and reveal the food.

A hungry puppy needs careful supervision if there is any tempting food left at nose height.

praise him periodically for staying there. At first you may need to tie him or barricade him into this area until he has acquired the correct habit. Alternatively, you could use a baby gate to keep him out of the room where you are eating. He will then be able to see you while you eat and will get used to lying down and waiting patiently for you to finish.

Settling down

Having a puppy that will settle down for short periods is useful when you do not want to involve him in what you are doing – for example, when you take him visiting at a friend's house or when visitors come to see you.

Begin in your own house when your puppy is slightly older, used to being restrained by a lead and has learned the 'Down' voice-cue. Exercise him first to use up any excess energy. Settle yourself comfortably with a book to read, or a similar activity. Bring your puppy, on a lead, to your side and ask him to lie down. Lure him into the down position if necessary and put your foot on the lead so that he cannot move too far away. Praise him gently and relax. If he tries to get up, the lead will stop him going very far and he will soon lie down again. Praise him whenever he does so, but otherwise ignore him. Interrupt him if he chews the lead or your shoes, and then praise him for lying quietly. Giving him a chew will help him to settle down and be calm.

Begin with just a few minutes and gradually increase this until your puppy can lie still for about half an hour while you are engaged in something else. When he is familiar with the exercise, teach him to settle down in situations away from home in other people's houses, on public transport or anywhere your dog should learn to lie down peacefully.

Car travel

Being in a car is not a natural experience for a young puppy, so it is important to acclimatize him slowly. Learning to accept being on the moving floor of a vehicle as it twists and turns is an important lesson for pet dogs. A dog that accepts car travel as part of life can be taken anywhere and included in all aspects of his owners' lives.

On the very first journey home, it is best if your puppy travels on someone's lap or in a small cardboard box at their feet. At this time, he has just left his mother and litter-mates and this will be traumatic enough, without having to adjust to the car as well. During all subsequent journeys he should be put in the place where you intend him to travel when he is older (and bigger). Choose somewhere he will be safe and where his movement will be restricted, so that he does not learn to jump around and distract

Teach your puppy to settle down when requested by practising this technique often at home and elsewhere while he is still young.

the driver. If he is to travel on the seat, use a car harness so that he cannot fall off. Good use can be made of a travel cage to ensure that he remains safe, but first get him used to being confined in it at home, and make sure it is big enough for him to stand up, turn round and lie down, changing the size if necessary as he grows.

As soon as your puppy has settled into your household, begin to accustom him to the car by placing him in it to sleep for half an hour daily. Cover the entire area with soft, absorbent material, put him in and allow him to explore. Settle him in and close the door, being careful not to frighten him by slamming it or the tailgate too hard. Ensure he is sleepy and has had a chance to go to the toilet before doing so. Do not do this on a warm day, or if the car is in the sun or in a place where people pass by.

When he accepts the car as a place to settle down, begin to take him on very short journeys, sometimes with a walk in the middle as a reward. Negotiate corners and bumps carefully to avoid unsettling him. Remember that he cannot see out and so cannot predict what the floor he is sitting on will do next. If it is bouncing around wildly, he will become afraid of going in the car, which is the opposite of what you

Key points

- Place your puppy on soft, non-slip, absorbent bedding for the journey.

- Close the doors carefully without slamming them, and don't start the car until the puppy is shut inside, as the exhaust can be smelly, unpleasant and frightening.

- Drive considerately; remember that your puppy cannot see where the car is going, is not supported by a seat and cannot predict when the next corner will be coming up.

- Take corners, bumps and rumble strips at a slow speed and accelerate/decelerate smoothly; corners and bends cause the car to move unpredictably from your puppy's point of view.

Let your puppy get used to the car by encouraging him to spend time there. He needs to get used to car travel early in his life so he can be taken anywhere without fuss.

A travel cage may help a puppy feel more secure, providing you choose one that is the correct size for the breed.

intend. When you reach home, take him straight to his toileting area before you both go inside.

Continue in this way, gradually increasing the distance, until he can accept quite long journeys. By doing this, you will slowly accustom your puppy to car travel without frightening or exciting him. He will learn to view journeys in the car as nothing out of the ordinary and will wait patiently until you reach your destination.

Take care!

- Do not leave your puppy in a car on a warm or sunny day; temperatures inside a car can rise rapidly and puppies can easily get over-heated and die quickly of heatstroke.

If your puppy dribbles, drools or is car sick, it is likely that he is anxious and worried by the movement. To help him overcome this, take him on many short journeys, never going further than he can cope with at one time without stopping. For some puppies, this may mean taking them repeatedly to the end of the road and then driving slowly home. Gradually accustoming your puppy to travelling in this way may take a long time at first, but it will be well worth it in the end as you will have a dog that is happy to travel and can be taken anywhere.

Before you let your puppy out of the car, insist that he waits for a while after the door or tailgate is opened and praise him for doing so. Restrain him if necessary. Once he knows that car journeys often end in a walk, he will be anxious to get out. Make this a habit while he is a puppy and you will have an adult dog that is under control when you open the car door, and that sits politely waiting until you release him before he jumps out of the car.

CHAPTER 14
Learning to be alone

Dogs are sociable animals and it is not natural for them to be isolated from others. Puppies need to be patiently taught to tolerate isolation slowly and in a structured way if they are to be completely at ease with it later on.

If this is not done and the puppy suddenly finds himself alone for hours, much upset and anxiety are caused which will occur again throughout his life whenever the puppy is isolated. A dog that has not learned to cope with being alone may scratch at doors, dig at carpets, pace endlessly, knock things off surfaces and windowsills, bark, howl and lose toilet control. It is easy to prevent this by slowly teaching your puppy to cope with being alone, especially if you start from an early age. Just like the process of socializing, you simply need to make the time to do it.

How to do it

Puppies, like all young, helpless animals, fear abandonment by their parent figure until they mature and become more self-reliant. Since you have become a substitute for their mother, you need to teach your puppy gradually to be independent in a manner similar to the way it would happen naturally.

As soon as you get your puppy, begin the process of teaching him to accept being alone.

Your puppy will need to learn to cope with the feeling of loneliness that he experiences as soon as the door between you closes.

Choose a time when he is tiring and is likely to settle down for a sleep. Play with him a little and take him outside in case he needs to go to the toilet. Then put him in his bed and shut him in the room alone. Puppies will often feel safer if they have a den-like area to sleep in. Putting his bed under a table or into a recess, or in an indoor kennel with a blanket draped over it, may help him settle more quickly.

Ignore any whining, barking or scratching at the door. If he is tired, he will soon accept being on his own and settle down to sleep. While he is very young, open the door once he is asleep. He can then come to you when he wakes up and you can take him out to go to the toilet. Repeat this exercise many times, building up the time that your puppy spends on his own until he can cope easily with a few hours of separation.

As well as actively teaching him to cope alone, make a practice of sometimes shutting doors behind you so that he cannot follow you from room to room. If you do this particularly when you are quickly going back into the room he is in, he will learn that you will be back soon, so there is no need to worry.

Never go in to a puppy that is making a fuss. If you do so, you will be rewarding this behaviour and he will do it more next time. Wait until your puppy is quiet before you enter, then go in and praise this behaviour instead. Go in as soon as there is a quiet moment; leaving your puppy to cry for hours on end will only make him fearful of being left alone. Build up to longer absences gradually, but never faster than your puppy can cope with. If he has a very gentle and clingy nature, you will need to make more of an effort to teach him to be left alone than you will a puppy of more independent character.

Never punish a dog when you return after an absence, no matter what has happened while you have been away. Your dog will not be able to link the punishment with what he did a long time ago, and it will not prevent him from doing it next time. He will, however, think that you are cross simply because you have returned. This will cause him to be anxious next time you leave him, since he will now be worried about you coming back and this may cause separation problems later.

CHAPTER 15

Training your puppy

Dogs that respond to their owners and can be controlled by a signal or a single voice-cue are much easier to live with. They probably have a better life as a result. Having to shout at your dog continually to get him to do what you want, or physically restrain or manoeuvre him, is hard work. It is not pleasant for the dog, either.

The object of training is to provide your puppy with a set of voice-cues that he understands, which can be used to make everyday life with him easier. A by-product of training is that you have a dog that is more in tune with you and begins to anticipate what you want so that he can be rewarded for it.

Using training methods that rely on reward for doing the right thing, rather than on punishment for getting it wrong, means that you can teach a puppy at any age and as soon as he has settled in.

Trial, success and error

Puppies, like us, learn by their successes and their failures. If a dog burns his nose on a fire, he learns not to do it again. If a dog barks for attention and everyone ignores him, he will eventually give up. But if he does something that is rewarded, such as raiding the bin and finding food inside, he is likely to repeat it.

Repeating rewarding behaviour, or ceasing to do things that are unpleasant or unrewarding, is the basis of any learning experience. This is true as much for humans and other animals as it is for dogs. In

Training your puppy with positive methods that use food, play and praise will strengthen the bond between you.

order to train a puppy, you simply need to make use of this process and manipulate situations so that he learns what you want him to do.

The correct relationship

Before you can begin training, you must have the right relationship with your puppy. He must view you both as a leader and as a friend.

Successful training depends on being seen as important enough to issue instructions. If your puppy looks on you as an equal or, worse, a weaker member of the pack, he will think you have no right to tell him what to do. By becoming a leader, what you want will become more important than what he wants. The higher your perceived status, the harder your puppy

will try to please. This high status should be achieved through everyday dealings with your puppy (see page 78); you should not wait until the time comes for training to show him who is boss. Otherwise force is needed, which will make your puppy either resentful or apprehensive.

Another essential element for reward-based training is that the dog should see you as a friend. If he does, he will be relaxed in your company and there will be no apprehension to inhibit the learning process. And being your friend will make him want to please you because he appreciates your praise when he does something right.

The old method of training involved a considerable amount of punishment, often administered by means

There are many different types of food treat that can be used for training. Soft, smelly treats that are easily handled are most effective.

of a choke-chain. Strong-willed trainers used punishment for not getting it right. This method is very unpleasant for the dog, which is why puppies were once left untrained until they were six months old; before this age, puppies were too young to take the rigours of this method.

Motivation and reward

All animals learn more quickly, and retain more of what they have learned, if they are not fearful. Consequently teaching using only rewards is far more effective and allows the puppy to put all his energies into learning the task in hand. An atmosphere of trust between puppy and trainer allows him to be more creative, and he is more likely to try new ways to obtain the reward without fear of being wrong.

A reward can be anything that the puppy wants. The most obvious and easiest rewards to use are:

• Food
• Pleasant social contact with humans
• Games with toys.

Different dogs are motivated by different things, and you need to find out what is most rewarding for your puppy. Reward-based training relies on the puppy wanting the rewards you are offering sufficiently for him to put himself out to obtain them. Having a reward that produces enough motivation is important, because the more motivated the puppy is, the harder he will try to learn the task.

Before you begin a training session, make sure you are using the rewards your puppy wants at that particular time in his life. He may be hungry just before a meal and so food will work well. Or he may not be hungry, but may be feeling playful, so a game with a toy may be a good reward. Skilful reward-based training requires you to correctly guess what your puppy would really like at that moment. Alternatively, you can manipulate the situation so that you cause him to want what you are offering at the time of the training session.

Grade the rewards you are offering according to what your puppy likes most, and use better rewards for tasks that are more difficult. The best rewards should be saved for things that are very difficult, such as coming when called from playing with children and other puppies.

After a few weeks of training sessions you may find that your puppy loses interest in the rewards that he used to work very hard for. Like us, puppies can get bored with the same old thing, and changing the rewards on a weekly basis will bring the response back up to your expectations.

If you are struggling to keep your puppy's attention and his motivation is low, increase the value of the rewards or change the type of reward you are offering. Also consider other internal motivations that may be preventing him from learning, such as the need to go to the toilet, being thirsty or feeling tired or unwell.

Food rewards

If your puppy has just eaten and you are offering food treats, he is not likely to work very hard to earn them. However, this does not mean that he has to be very hungry when you want to train him. In fact, being

A relationship that is founded on trust and respect is the best basis for training.

hungry could actually disrupt the learning processes, as he may be able to think of little else other than obtaining the food. Just before a meal is the ideal time for training, when your puppy is peckish, but not ravenous.

Your puppy will work best for food treats that are smelly, soft and tasty. Cooked sausage or liver that has been cut into small pieces works well. Drying it in the oven means it loses its softness, but it can then be stored in an airtight container so that you always have a ready supply to hand. Keeping several

containers in different rooms of the house ensures that you are never far away from the rewards, should a few minutes of your time become available for a training session.

Whether you decide to use home-made treats, one of the many dog treats on the market or something else, make sure the pieces are small: about 5 mm (¼ in) cubed or less (about the size of a large pea). Strangely enough, dogs seem to work harder if the pieces are small. They also become full less quickly, so the training sessions can last longer.

If you are worried that your puppy may get fat with all the food treats used for training, reduce his meals accordingly. When you first start training, introduce the food treats you will use slowly, to avoid upsetting his stomach with large quantities of a new food, a little at first and a bit more each day.

Pleasing you

Using food when training puppies makes it easier for them to understand what it is you want. Although they are rewarded by the food, they will also be looking for your approval. Puppies will learn much faster if you reward them, not only with the food, but with happy praise and genuine approval whenever they do what is required.

Verbal praise is most effective if given in a high-pitched, happy, excited voice, and physical praise is best delivered by stroking the back and chest areas gently and lightly (avoid patting the head – how would you like it?).

If you have the correct relationship with your puppy, trying to please you will, eventually, become a much bigger motivation for him than working for food rewards alone.

Toys and games

Later, when your puppy has learned the basic voice-cues, you can use games with toys to motivate him to do as you ask. Games with toys are a continuation of your approval, and having fun with you in a game is highly rewarding for dogs once they have learned how to play with toys (see Chapter 9, page 92).

Keep pots of treats in convenient places in different rooms of the house to allow spontaneous training sessions whenever you have a minute spare.

Teaching your puppy to sit in the living room is easy to do because you can practise it often.

Some puppies are not very motivated by food, but will work for toys instead. Remember that the reward for your puppy is the game, and not just the winning of the toy, so make sure you play with him for a short time when he has got something right. This, together with getting the toy back from your puppy at the end of the game, takes time, so it is more convenient to use food treats to train at first, if you can.

Different places

If you teach your puppy something in one place (such as to sit in the living room), he will not know how to do it when you say the word in another place (such as if you ask him to sit in the garden or at the kerb). This is because puppies learn a whole set of associations surrounding an event, not just the one thing you are trying to teach.

Say, for example, you teach your puppy to lie down when asked, while you are sitting on a chair in your living room. Your puppy will probably have picked up that:

- If you are in a certain position in that room and
- You are seated in a particular chair and
- He is facing you and
- You give the voice-cue in a certain tone of voice, he can get a reward for lying on the floor.

He will not have learned the rule 'When my owner says "Down", if I put my elbows on the floor, I get a reward.' Take away any of these associations and he may have difficulty in understanding what you want. So if, for example, you take him out to the kitchen, stand beside him and ask him to lie down, he will not know what you want him to do. This will be so, even

Teaching your puppy the same exercise in different locations is important if he is to learn to perform on the signal only.

though he is more than willing to do something in order to get the reward.

To solve this problem, you need to remove all of the associations one by one (apart from the voice-cue itself) by teaching him all the exercises in many different locations, with him in many different positions in relation to you, using different tones of voice. By doing this, he will learn that it is the voice-cue itself which means that a reward is available for doing the required action. (For most pet dogs, 'Sit' is the only voice-cue that they ever learn, because it is the only one that owners repeat many times in many different situations.)

Practise each exercise in many different places. Do not be surprised if, in the early stages, your puppy appears not to understand a voice-cue that he is very familiar with when he is in a different area. Simply show him what it is you want, by patiently going

through the teaching process again in that situation, and he will soon learn to respond to the voice-cues you give. Every time you do this you will be adding to your puppy's understanding and he will take less time to work out what you want next time. Eventually he will respond the first time, wherever you happen to be.

Practise whenever you can, and particularly when it is also useful to you to have him take up a certain position, such as asking him to 'Stand' so that you can dry him underneath with a towel if he is wet from a walk or bath. If he doesn't respond, give him a few moments to think about what he needs to do, then use a food lure to get him into position. Remember to praise him well and give a treat when he does the right thing.

If you want to test whether your puppy really knows your voice-cues, get someone to hold his lead, turn your back on him and give your voice-cue. If he genuinely knows the words, he will do as you ask. If he just stands there wagging his tail, he does not know them yet.

Incorporating distractions

Once your puppy is responding to your voice-cue or hand signal in lots of different places, begin to teach him to respond when there are other things going on around him. Start with distractions that are not too exciting and gradually work towards things that your puppy may prefer to do instead of what you are asking. Make sure you have a really good reward if he responds, so that he realizes that it was worth it, and he will be more likely to do so again next time.

Random rewards and jackpots

One of the biggest objections to using rewards for training is that you always need pockets bulging full of treats and toys. However, it is only necessary to reward continuously when your puppy is in the early stages of learning. Once he has learned what to do, you can considerably reduce the rewards given. This may seem strange, but scientific experiments have shown that animals will work harder if they are rewarded occasionally at random, rather than constantly.

So once your puppy has learned what a voice-cue or hand signal means, there is no need to reward every response. This means rewarding occasionally – twice in every ten times at random. The transition between rewarding constantly and variably should be

Providing occasional reward jackpots, made up of extra-special praise, food and games, will make your puppy work harder to try to win them.

Key points

- Only begin rewarding at random once your puppy has fully understood what it is you want him to do. Randomly rewarding before he has learned the exercise will confuse him greatly and it will take longer for him to learn.

- Complicated tasks or difficult requests should be rewarded every time. If your puppy finds something difficult (like a complex trick), or doesn't want to do something but responds anyway (such as coming away from puppy play when you call), make sure you reward every time. Random rewards won't work in this context, as the effort required to do as you ask is not outweighed by the 'fun' of gambolling. Reward well and be pleased that your puppy made such an effort for you.

done gradually; for example, reward eight out of every ten times at first, then six out of ten times, then five out of ten times, until you reach twice in every ten. Reward only the better responses and you will encourage these rather than the weaker ones. Sometimes, especially if the response has been slow, offer no reward except a word or two of praise to let your dog know that he has done the correct thing.

Random rewards can be a difficult concept to put into practice because you will be used to rewarding constantly and may feel mean about not rewarding. However, don't forget that you can still praise your puppy to let him know he has done the right thing. If you do this, there are great benefits, as your puppy cannot anticipate when he will get the reward and so will work longer and harder for the same amount of reward than if rewarded continuously.

To make training even more fun for your puppy, and to make it even more likely that he will respond,

offer occasional jackpots. At random, every now and again (about once in every 20 responses), offer a 'jackpot' of praise, food and games. These are a special treat, or abundance of treats, or several really nice things at once. Jackpots need to be extra special, so have fun, dance around and celebrate with your puppy! It is like winning the lottery, and he will begin to gamble on the outcome and continue to respond 'just in case'. If you save jackpots and reinforcement for the best performance, his performance will improve.

You will find that, using this system, your dog will perform much better than if he were being rewarded every time, since he will hope that the next time he will get the jackpot. He is, in effect, gambling on the outcome – putting energy into a system where sometimes he may get nothing, but sometimes he may receive a huge reward. This system works with people too, and it is the reason why people gamble on horses or play the lottery. Dogs like to 'gamble' too, and you can use this to good effect in your training.

Timing

Good timing is essential for effective training. Rewards need to be given as soon as your puppy attempts to do the right thing so that the required action is rewarded and encouraged. Immediate reward is best, but any time up to two seconds after the required action will do.

Rewarding too late brings confusion because something else is being rewarded. Say that you encourage your dog to go into the 'sit' position and take one minute trying to get a food treat out from the bottom of your pocket. By the time you reward him, your dog will have mentally moved on to thinking about something else (perhaps getting up), even if he is still sitting, and it is this that you will be rewarding rather than the act of sitting, which is what you intended to reward.

Good timing is not difficult, but it is a skill that needs to be learned. Some people seem to have a natural ability, but it is probable that they have just had more chance to practise. Playing the training game (the original idea for this type of game came from leading dolphin trainer Karen Pryor, author of *Don't Shoot the Dog*) will allow you to practise this essential skill before attempting to train your puppy,

Taking too long to produce the reward will mean that your puppy moves on to thinking of other things and the chance to reward the behaviour is lost.

thereby preventing your lack of skill from confusing him and getting you both off to a frustrating start.

The training game

You need at least two people. One person elects to be the 'trainer', the other is the 'dog'. The 'trainer' stays in the room and thinks up a task while the 'dog' goes out until invited in (keep the task simple to begin with; for example, the 'dog' has to come in, sit on an appointed chair and put their hands on their head).

The 'trainer' has to help the 'dog' to do the task without anyone speaking. The 'trainer' is allowed to reward the 'dog' with a blast on a whistle (or a hand clap if you do not have a whistle) whenever the 'dog' progresses some way towards the desired goal – for

example, takes a step in the right direction or produces a behaviour that is something like the one required, such as putting their hands up to their ears. The 'dog' will need to be as inventive as possible so that he can give the 'trainer' something that he can reward.

The 'dog' comes into the room and tries different movements to get a reward from the 'trainer'. He will eventually be successful and the reward from the 'trainer' will ensure that the 'dog' builds on this progress until the task is eventually completed.

Being both 'dog' and 'trainer' in this game is a useful experience. As 'trainer' you will find out how critical accurate timing is, and how the 'dog' keeps repeating actions that are rewarded accidentally or because the reward came too late. Do not blame the

Puppies tire easily so keep the lessons short and allow plenty of time for your young puppy to rest.

'dog' for this! Practise being the 'trainer' until your reactions are sharper. As a 'trainer' or observer, you will also realize how long it takes to get the required action and how difficult it appears for the 'dog' to understand what is required.

As the 'dog', you will find out how frustrating it is to work for a 'trainer' whose timing is slightly out. Even if the timing is perfect, it will still be difficult to understand what it is that the 'trainer' requires. Finding out how taxing it is to be the 'dog', even if you have an experienced 'trainer', is a useful and humbling exercise.

As the 'dog', you will realize how much pressure there is on you to do the right thing, particularly if there are a number of people in the room. Puppies

feel that pressure, too, especially as they become more 'educated' and realize that you want them to do something in particular which they cannot immediately grasp. This is why you will often see them sit down and scratch or yawn when confused; they are engaging in displacement activity to relieve some of the tension they are feeling.

By being the 'dog', you will also learn how tiring it is to have to concentrate on what is required – a good reason for keeping training sessions very short. It is also useful to ask the 'dog' what he learned at the end of the game. Often the 'dog' will have picked up many actions associated with the task that you did not intend him to learn. For example, you may have intended him to learn to come into the room, sit on a particular chair and put his hands on his head, but he may actually have learned to come into the room, walk towards the fireplace, turn round twice, then sit on the chair and put his hands on his head instead. These unnecessary actions, which are extra to those required, are called 'superstitious' behaviours, because the dog thinks it needs them in order to get the reward. It is useful to realize how easy it is to teach dogs 'superstitious' behaviours during training.

Teaching and learning without the benefit of speaking the same language is not easy, but it is possible. By playing the training game, you will get to know how it feels to be your puppy and will also learn (or improve on) skills that will help you with your training. Reading how to do it is not enough. In order to acquire the necessary skill, you must go away and play this game before you attempt training with a real puppy!

Keep lessons short

Learning is a tiring process and young puppies do not have the attention span, stamina or powers of concentration for long sessions. Keep lessons very short – about one to three minutes – but do lots of them throughout the day.

Is your puppy enjoying training?

How well your puppy learns will depend on how he is feeling. Dogs, like humans, learn best if they are

Immediate rewards, given as soon as your puppy does the right thing, are essential for successful training.

ACTION	VOICE CUE	BODY POSITION	HAND SIGNAL	ACTION	VOICE CUE	BODY POSITION	HAND SIGNAL
Give attention	'Watch me'			Roll over	'Over'		
Sit	'Sit'			Walk close	'Close'		
Lie down	'Down'			Come here	'Here'		
Stand up	'Stand'			Stay	'Stay'		

feeling well and happy. Feeling off-colour, frightened or tired will inhibit the learning process.

For some puppies, having too much physical energy can also have this effect, and a chance to run and play before a training session can improve things. Dogs learn more quickly if they are enjoying themselves, so keeping sessions fun will also speed up the process. Keep that tail wagging!

Body postures and words

Dogs have a communication system based on body postures rather than on language (see page 24). Consequently they are much better at learning via body and hand signals than spoken words and sounds.

Long before they learn the meaning of each voice-cue, puppies will often use our body postures and hand movements as visual clues to help them work out what it is we want. Therefore to help them with the relatively difficult task of learning words, it makes sense to decide not only on a series of spoken words, but also on a series of hand signals and body postures as well.

The whole family will need to agree on this list and learn it, so that no conflicting signals are being given to cause confusion. Put up a table like the one above somewhere where everyone can see it. Being consistent with the signals you use – both verbal and visual – will help your puppy to learn more quickly.

Initially, use both hand signals and voice-cues when teaching your puppy. Later, as his understanding grows, you can begin to dispense with the hand signals and use spoken words only.

Being consistent with the signals you give during training will help your puppy learn more quickly.

All voice-cues should be different, and no two voice-cues should sound similar. Avoid choosing voice-cues that have a sound similar to the puppy's name, such as 'Kit' and 'Sit'. Make a copy of this list and pin it up in a prominent position so that all the family can see it.

How long will it take?

It takes regular training sessions over a period of about six weeks for one voice-cue to be learned reliably. Several sessions per day for a few months, at times of the day when your puppy is most responsive to learning, will soon establish the basic voice-cues.

A training programme

Work out a training programme and set aside time each day until your puppy is at least 12 months old. Set achievable goals to keep yourself motivated. Include all family members in the programme, even young children (who will, of course, need supervision). It only takes a few minutes each time, so choose training times when you are not doing anything important, such as while the kettle is boiling.

One of the best times to train is when you are out on a walk. Since you have set aside that time for your puppy anyway, it makes sense to use it to your best advantage. Short play/training sessions every now and then as you walk along will result in a well-trained, responsive puppy that is obedient outside as well as at home.

If you work hard on your training programme for the first year of your puppy's life, you will have an adult dog that knows the basic voice-cues and will work well for you. All that will be needed after this is a little extra occasional training to refresh his memory. Throughout the training, ignore any of your puppy's actions that you do not want and reward those you do. Do not tell your puppy off if he gets it wrong, just work on encouraging the correct response and reward it well.

Aim for success

During your training you should aim to provide a situation where your dog gets things right first time and is rewarded for it. Do not allow situations to

Playing the tail-wagging game will help the training process by teaching your puppy to watch you for clues about what to do next.

develop where you are repeating your voice-cues and your puppy is confused or taking no notice. If you do this, you will rapidly desensitize him to the sound of your voice and he will learn to ignore it.

If you reach a point where you are not progressing, stop and think about why you cannot proceed. Plan your training sessions so that your puppy learns something – however small – every time and can be rewarded for it. If the new exercise is not going well, go back to something he knows how to do, so that you can praise him for it and end the session on a high note.

Practical exercises

Young puppies have a very limited attention span, so keep lessons short and simple at first. Training should be fun for you and your puppy, so make the training sessions informal and break off into a game at any time to keep learning exciting.

Tail-wagging game

Before you can teach your puppy something, his attention must be on you. You cannot teach anything if your puppy is looking the other way, walking away from you or sniffing the floor around your feet. Before beginning a training session, play this tail-wagging game.

In a place where your puppy is not distracted by anything else and when he is looking elsewhere, go up quite close and say his name, quietly and clearly. As the puppy looks up at you, reward him instantly with a small food treat and lots of praise. Let him see that you have another treat and hold it under your chin. Say his name, use your voice to excite him and make his tail wag as much as you can. When he is at his most excited, give him the treat.

Repeat the game often in many places or situations until your puppy will give you his undivided attention instantly whenever he hears his name. You could have a contest among family members to see who can make his tail wag the most.

Then gradually begin to hold his attention for longer by quietly talking to him and maintaining eye contact before giving the treat. This helps to build his concentration time which will make training easier, as will occasionally playing this game in more distracting circumstances.

Come when called

Practise steps 1–3 on the following pages. You will not always need someone to hold your puppy while you call him. Just wait until he is some distance away, and thinking of something else before calling him excitedly and doing the rest of the exercise. You need to make him want to come to you, not just for the treat, but because he wants to be with you. For this reason it is really important that you connect with him mentally and let him know how clever you think he is when he arrives. Don't just feed the treat and leave it at that.

Only call him when you can be reasonably sure of success. If you call your puppy when he is far too excited about something else to respond – for example, when greeting one of the family who has just come home – you will be teaching him to ignore you. In the early stages, never call him at such a moment. Wait until the excitement has subsided and he can be distracted, before calling him.

He needs to get into the habit of coming to you immediately he hears his name. You should aim to call him back to you successfully as often as you can.

Recall training will allow you to be confident about letting your puppy off-lead to have more freedom and a better life.

Take every opportunity to call him when he would be coming to you anyway; for example, when he notices you have his dinner or when you return to him after being away.

Always have a really good reward ready for him when he comes to you. And always greet him enthusiastically and be really pleased with him when he reaches you. You will need to be as enthusiastic about him coming to you each time as you were the first time he did it (make that tail wag). Once he has learned the voice-cue, reward only the more rapid and enthusiastic responses with food treats or games. If this continues throughout his puppyhood, you will have a very strong response by the time he is older.

When your puppy rushes to you instantly every time he hears his name together with the voice-cue to come, you will need to teach this exercise in different places and in more distracting circumstances.

During a walk, call your puppy back occasionally, reward him well with food treats, a game and praise

Exercise: Come when called

▷ **1** Ask someone to restrain your puppy and release him as soon as you call. Go to your puppy and show him that you have a tasty food treat (something he really wants) by holding it in front of his nose; but don't let him take it! Run backwards a few paces (being careful not to trip over), stop and call him to you, calling his name and your voice-cue in an exciting, encouraging way.

▷ **2** Be 'open' with your body language by holding your arms outstretched. As he begins to run towards you, praise him enthusiastically and continue to call him excitedly.

for coming, and then let him free again. It is important to do this during your walk, because otherwise you may be calling him back only at the end of the walk when you want to put the lead on to go home. Your puppy will learn that being called means the end of his freedom and he may avoid you.

Never spoil a good recall by punishing your puppy for coming back. It is too late to be cross about whatever he has done. The important thing is that he did, eventually, come back. Punishing him when he returns only makes it less likely that he will come back next time. And never call your puppy when you want to do something that he may not like, such as confining him or giving him a bath. Go and fetch him instead.

Walking on a lead without pulling

While your puppy is still young and small, it is important that he learns that pulling on the lead means that he stops rather than goes forward. Most

▷ **3** When he gets to you, hold the treat out to him with one hand, with your other hand outstretched below it. As he moves forward to take the treat, gently draw him towards you, so that you take hold of the collar underneath his chin without reaching forward. Once you have hold of the collar, release the treat. Praise him warmly. Keep hold of his collar until he has accepted the restraint, then release him and allow him to wander away. Do not reach out to grab him. If you do, he will learn to avoid you. Do not put your hands all over his head as he will probably not like this and will, again, learn to avoid you.

puppies learn the opposite of this, which is why so many dogs pull their owners along the street. Puppies that learn not to pull grow into dogs that are a joy to take out. Because of this, they are likely to be taken out more often and for longer walks.

To teach your dog to walk nicely on the lead, you will need to be consistent and *never* allow him to pull, from the first moment you clip the lead on. If dogs are tied to a post, they learn that it is not worth pulling. You need to teach your puppy that if he pulls on the lead, you turn into a post that cannot be pulled. This takes time and needs much patience at first, but the rewards will be worth it when your puppy gets older and larger.

Teach your puppy to walk on a loose lead as soon as he has got used to his collar and no longer takes any notice of it being around his neck or being

Exercise: Walking on a lead without pulling

▷ **1** Play an energetic game with your puppy to use up any excess energy, then attach a lead to his collar and stand still. Hold the lead in the right hand, with the food treat in the left hand nearest your puppy. The lead is there just to stop him moving a long way from you, so anchor it by pressing your right hand against your body and try not to move it from this position. Don't be tempted to use it to pull him towards you. Adjust the length of the lead so that it is loose, but not so loose that it drags on the ground. Encourage your puppy to stand somewhere close to your left leg by luring him there with a food treat (keep your feet still). Praise him and feed him the treat when he gets there. Keep doing this until he stands still by your left leg waiting for the next treat.

▷ **2** Show your puppy you have another treat, stand up straight and hold the treat up out of his way. Say his name to attract his attention, give your voice-cue for walking on a loose lead and move forward one step.

restrained by it (see page 52). Always use a flat buckle-collar, never a choke-chain.

Repeat steps 1–7 of this exercise until he will walk beside you without pulling on the lead wherever you go. This takes patience and time at first, but improvements will be seen after a few sessions. Don't become disheartened if you do not see perfect results the first time as it is a complicated exercise

and it will take a while for both you and your puppy to learn how to do it well.

Practise this exercise in the house and garden while your puppy is still unprotected by vaccinations and cannot go out for a walk. He should have learned to walk calmly beside you, without pulling on the lead in these locations, before you begin to take him out to exciting places

3 Bend down, feed the treat and praise him warmly. Repeat steps 1–3 until you can do it easily.

4 As you begin to feel more confident and have your puppy in the correct position, begin to take more steps, building up slowly and gradually, with your puppy walking nicely with you.

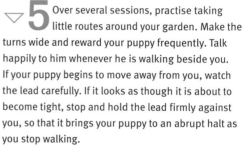

5 Over several sessions, practise taking little routes around your garden. Make the turns wide and reward your puppy frequently. Talk happily to him whenever he is walking beside you. If your puppy begins to move away from you, watch the lead carefully. If it looks as though it is about to become tight, stop and hold the lead firmly against you, so that it brings your puppy to an abrupt halt as you stop walking.

▶▶

Going for walks

When your puppy goes out for his first walks, he will be excited by all the new smells and sights and will want to rush forward to investigate. It is important to instil good habits straight away by never allowing him to walk along pulling on the lead. Be patient, and always allow enough time for walks at first. It will be slow to start with, but if you make a point of never allowing your puppy to pull on the lead, the teaching process will take less time overall. When you want to stand still, for example, to talk to someone, do not let your puppy pull you away from the place where you are standing. If you do, he will learn to pull into his collar in order to get somewhere else. If he can never go anywhere by pulling against his collar, he will learn that it is pointless to do so and will give up.

If you have to go on walks during which you are concentrating on something else, such as taking the children to school, it may be better to find an alternative way to walk your puppy until he is trained.

△ **6** Encourage your puppy back into position and reward him. Wait until he is calmly waiting next to you before repeating step 2.

Take care!
- Head-collars ensure that you have control of your dog's head and that he cannot pull. Be careful to use the lead gently if your puppy is wearing one – it is easy to injure his neck by jerking, pulling on the lead too roughly or if your puppy takes off after something and is suddenly brought to a standstill quickly when reaching the end of the lead.

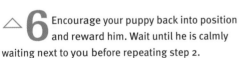

△ **7** As your puppy gains experience of walking on the lead without pulling, begin to incorporate turns and obstacles. Warn him first that you are about to turn, by getting his attention as you are walking along and encouraging him to come close to you as you walk round an obstacle. Make it fun to be walking with you by praising him warmly every now and again, and by feeding a treat after completing a turn or a change of pace.

Continue to practise, gradually building up the time in between treats. Don't forget occasional jackpots (see page 164). Talk to your puppy occasionally whenever he is walking close to you and praise him enthusiastically, to keep him with you. Varying the speed at which you walk will keep this exercise interesting. Speeding up suddenly will be more exciting for your puppy, and this can be used if he is becoming bored. Try to make his tail wag all the time when he is doing this exercise.

If he is small enough, you could carry him there and train him on the way home, or lead him with a head-collar, if you cannot train him. Allowing your puppy to pull on the lead during one walk, and then expecting him not to on the next, will confuse him and make it much harder to teach him to walk correctly.

Once your puppy has learned to walk on a loose lead, teach the exercise again in very distracting circumstances and when walking with other dogs. Your puppy will probably seem to have forgotten all he has learned when things get more exciting and interesting, but patiently go through the training process again as if you were starting from scratch. If you do this, he will learn that no matter where he is or what is going on around him, making an effort to keep near you so that the lead is loose will result in him getting somewhere faster than if he pulls. It is particularly difficult for him not to pull when walking with other dogs so arrange to go out with friends and their dogs so you can practise.

Key points

- If you hold the food treat in the wrong hand, your puppy will walk across in front of you to get it and will trip you up. Always hold the food treat in the hand nearest your puppy.

- Holding the food treat too low and close will encourage your puppy to jump up. Hold the treat up out of his way, and reward him with the treat when all four of his feet are on the ground.

- If your puppy goes round behind you or cuts across in front, stop and reposition him. Do not turn round as you do this – lure your puppy into position rather than moving yourself, so that he learns to manoeuvre himself relative to you.

- Professional dog handlers walk their dogs on the left, but there is no reason why you should not walk your puppy on the right. Just keep to one side only at first, and ensure that all the family does so, too.

Positions: sit, down, stand, roll over, stay and down at a distance

All 'positions' can be taught in the same way. The puppy is lured into position with a food treat, the voice-cue is given as the puppy goes into position and he is rewarded as soon as the required position is reached.

Remember that your puppy will learn body language and hand signals before the spoken word. Keep these consistent throughout his training and reduce them gradually as his response to your spoken voice-cue improves.

Sit

Practise steps 1–3 below. When your puppy sits as soon as the food treat is held above his nose, incorporate the voice-cue 'Sit'. Say it obviously and clearly, just as your puppy's bottom sinks towards the floor. Remember to feed the food treat and to praise warmly while he is in the sit position.

Exercise: Sit

▷ **1** While the puppy is standing, hold a food treat just above his nose. Hold the treat tightly between your thumb and forefinger so that he cannot get it until he has moved into the correct position.

Key points

- If he jumps up with his forepaws off the floor, the food treat is being held too high.

- If he backs up, the food treat is being held too far back and too low, or you are moving it back too fast.

- If you have chosen 'Sit' as your voice-cue, be careful not to say 'Sit down' by accident, especially if you have chosen 'Down' for the lying down position, as this will cause confusion for your puppy.

Practise this exercise in short sessions, until both you and your puppy have mastered it and your puppy is sitting down each time you raise the food treat above his head. During further sessions, keep the food treat out of sight, say your puppy's name to get his attention and to get him looking at you, ask him to 'Sit' while giving a clear hand signal, then bend down to lure him into position with the food treat. After many repetitions, you will find that he begins to learn the voice-cue and starts to sit without being lured into position. When he does, feed him the treat at once and praise him enthusiastically.

Once your puppy knows the voice-cue, begin again by teaching him in different positions in relation to you: when he is standing in front of you, standing beside you, when you are sitting down or standing up. Teach him to sit on voice-cue from a lying-down position. Teach him in different places and with distractions going on around him.

2 As the puppy raises his nose to take the food treat, move the treat up and back towards his tail, keeping it just above his nose all the time. As his nose and head move up and back, his bottom should automatically sink down towards the floor.

3 As soon as your puppy sits, feed the treat and praise him well. Stroke him gently and continue to praise him warmly and to feed him treats for as long as he remains in that position.

Down

This is the most difficult position of all to achieve and you will need perseverance and patience. Try this exercise just before your puppy's next meal so he is hungry and use a tasty treat that he will really want to work for. Practise steps 1–4 below until your puppy understands what 'Down' means.

Once your puppy knows the voice-cue, begin again by teaching him in different positions in relation to you: when he is sitting in front of you, sitting beside you, when you are sitting down or standing up. Teach him to lie down on voice-cue from a standing position. Teach him in different places and also with distractions going on around him.

Exercise: Down

△ **1** With your puppy in the sit position, use a food treat to lure his nose down slowly to the floor. Hold on to the treat if he tries to take it from between your fingers, but allow him to chew small pieces off the edges to keep him interested.

△ **2** Use the treat to keep his head as close to the floor as possible. As his head is now down, it will be easier for him to lie down than sit hunched over. Wait patiently for this to happen. Be calm and wait without saying anything to distract him. If he stands up instead of lying down, lure him back into the sit position, feed a treat to reward him and then try again.

▷ **3** As soon as his elbows reach the floor, feed the food treat and praise him well. Stroke him gently and continue to praise him warmly while he stays in position. When he is lying down easily, begin to incorporate the voice-cue by saying 'Down', obviously and clearly, just as his elbows sink towards the floor, before rewarding him well.

Key points

- If he keeps standing up, try holding the treat closer to his paws so that he doesn't have to move forward to reach it.

- Many puppies do not like to lie down on a cold or hard floor. If it is difficult to get your puppy to lie down, try again on a piece of soft bedding.

- If you are not having success, try rewarding him for lowering his head at first, then ask for a little bit more, and then a little bit more, until his head is right down and it is easier for him to lie down to get the treat.

- If it is very difficult to get your puppy to go down, make a bridge with your legs and try luring him underneath. Feed him several treats in the down position so that he learns this is what you want, then swivel your legs away and try again with your puppy in the same place.

- If you've chosen 'Down' as your voice-cue for this exercise, try not to use it at other times. For example, it is tempting to say 'Get down!' when your puppy jumps up or gets on the sofa. If you use 'Down' at these times you will confuse him and weaken his response to the voice-cue.

△ **4** Practise stages 1–3 until he is lying down easily, and then teach him to do so with a hand signal and voice-cue instead of the lure. While he is in the sit position, keep the food treat out of sight, get his attention and ask him to lie 'Down' while giving a clear hand signal. Wait a few moments and then bend down to lure him into position with the food treat. After many repetitions, you will find that he will begin to learn the hand signal and voice-cue and to lie down without being lured into position. When he does, feed the food treat at once and praise enthusiastically.

Stand

Practise steps 1–4 below until your puppy understands what 'Stand' means.

Once he knows the voice-cue, begin again by teaching him in different positions in relation to you: when sitting in front of you, sitting beside you, when you are sitting down or standing up. Teach him to stand on voice-cue from the down position. Teach him in different places and also with distractions going on around him.

Exercise: Stand

▷ **1** With your puppy in the sit position, put a food treat just in front of him and, keeping it level with his nose, draw it slowly away.

> ### Key points
>
> • If you move the food treat away too fast, he may just sit still, waiting to be rewarded in that position. If this happens, move the food treat back and go more slowly.
>
> • If the food treat is moving too fast when he gets up, he may walk forward, and will learn to do this rather than just standing up, which is what you want him to learn.

△ **2** As he moves forward to follow the food treat, he should automatically stand up.

△ **3** As soon as he is on his feet, stop the food treat moving forward, feed the treat and praise him enthusiastically. Stroke him and continue to praise him warmly while he stays in position.

▷ **4** After several sessions of practising stages 1–3, when your puppy will stand easily when lured, teach him to do so when you say 'Stand'. With your puppy sitting, keep the food treat out of sight, say your puppy's name to get his attention, ask him to 'Stand' while giving a clear hand signal, then bend down to lure him into position with the food treat. After many repetitions you will find that he begins to learn the voice-cue and starts to get up without being lured into position. When he does, feed the food treat at once and praise him enthusiastically.

Roll over

This exercise can be very useful later on when your dog is fully grown and the veterinary surgeon needs to examine underneath him. Wrestling any dog to the ground so that you can look underneath is much more difficult than simply teaching him to roll over.

Practise steps 1–5 below. Once your puppy knows the voice-cue, begin teaching him in different positions in relation to you: when he is standing in front of you, standing beside you, when you are sitting down, standing up, and so on. Teach him in different places and also with distractions going on around him.

Exercise: Roll over

▷ **1** With your puppy in the down position on a carpet or mat, put a food treat just in front of his nose and bring it round slowly towards his shoulder.

▽ **2** As the puppy turns his head to get the food treat, his body should roll over to the side. Be patient, because it takes time and perseverance to get the hang of this.

◁ **3** As soon as he is on his side, praise him well and feed the food treat. Stroke him and continue to praise him warmly while he stays in position.

▷**4** Continue to do this for the next few days, gradually bringing the food treat round further before rewarding him so that, eventually, the puppy is rolling over onto his back as he follows your hand.

▷**5** After several sessions of practising steps 1–4, teach him to do it when you say 'Roll over'. With your puppy in the down position, keep the food treat out of sight, say his name to get his attention, ask him to 'Roll over' while giving the hand signal, then bend down to lure him into position with the treat. After many repetitions, you will find he learns the signal and begins to roll over without being lured into position. When he does, reward him enthusiastically.

Key point

- If you are struggling to get your puppy to roll over, try rewarding him for very small movements of his head to the side. Ask for a little more next time before rewarding him, and give him time to make the movement, rewarding him well when he does so. Also, try to position your hand so that you can sweep cleanly round his head to the other side, so that you do not have to distract him with a jerky movement.

Stay

The stay is a useful exercise for the many occasions when you want your puppy to stay put while you do something else. Having a word that means 'stay where you are until I come back to you' can be used, for example, when you want to leave your muddy puppy on the back doorstep while you fetch a towel from inside the house or would like your puppy to wait in the hall while you answer the door to take in a parcel being delivered. It is an easy exercise to teach, but teach it when your puppy is a little tired rather than full of energy.

Exercise: Stay

◁ **1** Ask your puppy to go into the sit position and praise him. Give your voice-cue and a slow hand signal to stay and keep still.

▽ **2** Count to two and reward him with a food treat and gentle praise while he stays in position (excited praise will cause him to get up and move). Stop rewarding him if he moves out of position and ask him to sit down again.

Repeat steps 1–2, gradually and slowly building up the time your puppy stays sitting for. Continue until you can count to 20. If your puppy moves, go back a stage and work up to a longer count more slowly.

Take care!

- Never use this exercise to leave a dog without supervision in a potentially dangerous situation, such as outside a shop. He may get distracted and forget his training; for example, if he sees another family member on the other side of the road, he may run across.

Slowly build up this exercise until you can walk past and around your puppy while he stays in position. When you have a reliable sit-stay, repeat this exercise in a place where there are distractions. You may also like to teach the down-stay, which will be easy if you have a reliable sit-stay.

Down at a distance

This is a useful exercise to teach because it gives you some distance control which may be useful in an emergency – for example, if your dog has wandered across to the other side of the road and there is no time to call him back before a car comes past.

△ **3** When your puppy can stay still in front of you for some time without moving, repeat step 1, but take one pace sideways from him.

△ **4** Once he stays reliably during the sideways movement, take two paces sideways and, watching your puppy carefully, step round beside him. Retrace your steps to the front and praise him while he stays in position. Gradually build this up until he can remain in position while you walk around him.

Obviously it is better to avoid such situations, but should they happen accidentally, being able to get your dog to lie down immediately might save his life.

Begin this exercise only when he has learned the 'Down' voice-cue and will reliably lie down in front of you as soon as he hears it.

Practise steps 1–3 below. Once he is reliably lying down at a distance, teach him to lie down as he is running towards you. Begin by saying 'down' when he has run almost to you. Gradually, over a number of sessions, stop him when he is further and further away, rewarding him well by throwing a toy or treat to him as soon as he lies down.

How to teach a trick

Tricks are fun to teach and, because you are enjoying yourself, your puppy will have fun too. All training (for example, 'Sit' and 'Stay') is really just a set of tricks:

Exercise: Down at a distance

△ **1** When your puppy is a pace or two away, give your voice-cue and hand signal for 'Down' as you step towards him. Reward him well as he responds.

△ **2** After a few repetitions, as he begins to respond more promptly, lean rather than step towards him.

your dog is doing something that you want him to do, in order to receive a reward and please you. Tricks can only be taught successfully by using the reward method so it is a good way to assess whether you are capable of teaching something without using compulsion. Tricks should be fun for both of you.

Make sure that the trick you choose is something that isn't demeaning or likely to cause injury. Do not teach young puppies to jump as this can injure their growing joints, ligaments and bones. Choose something that may be useful perhaps, such as walking backwards or carrying a tin of dog food. You will find it easier to teach the trick if you choose something your dog will naturally be good at: terriers like to use their paws, gundogs like to retrieve and hold things, and herding dogs like to chase.

▷ **3** Eventually you will be able to stand up straight, give your voice-cue and your puppy should lie down instantly a few paces away. Reward him well with a food treat, praise and a game if he responds correctly. Gradually, over a number of sessions, increase the distance between you before giving your voice-cue. If, at any time, he does not seem to understand, go back a step and repeat the exercise with him closer to you.

Always reward your puppy as soon as he does what you want. In the early phases of teaching each stage, reward him as soon as he attempts to do the right thing. Praise and give the food treat at once (make that tail wag). Don't ask for too much at once. Remember the training game (see page 165) and how disheartening it was when you did not receive any reinforcement, despite trying several things you thought would be right.

Never get cross – if your puppy will not do what you want, think again; you have made it too difficult. Keep to two-minute sessions only, then stop and ask

Exercise: Teaching the crawl

▷ **1** Begin with your puppy in the down position, placing a tasty treat against his nose.

▷ **2** Slowly move the treat forward, keeping it close to the ground so he has to crawl to get it.

▷ **3** After many repetitions, move your hand away faster and introduce a voice-cue and hand signal so that your puppy learns to 'crawl' without the lure.

The 'spin' is easy to teach by using a treat to lure your puppy round in a circle. Once he has learned this, teach a 'twirl' in the other direction too, to develop his muscles equally on both sides.

Teach your puppy to do a 'high five' by pawing at a raised hand.

yourself what your puppy has learned. If the answer is nothing, think about how to make things easier for him to do what you want. Always have a rest between each session. Practise each stage until, when you begin again, your puppy does the required action first time.

With complicated tricks, the secret is to break them down into small sections and then put these together to form a sequence. Always work on the last part of the trick first and then work back to the beginning. This ensures that something familiar follows on from each new stage you teach. For each small section, work out a way to position the reward so that your puppy can only obtain it when he has performed the action you require. Try to make it obvious to your puppy what he is expected to do so that he gets it right first time.

Once your puppy has learned the trick completely, don't forget to practise in different places and then with distractions. If he seems to have forgotten all he has learned when in a new place, start the training from the beginning again. He will learn much faster the second time, and faster still in your third new place.

Choose from these tricks or invent your own:

- Wave
- Walk backwards
- Walk through a hoop
- Be a courier between members of the family (go to Dad, Mum, etc.)

- Fetch the newspaper
- Carry the lead
- Carry your bag
- Take an object and drop it in the bin
- Shut the door.

Keep all training sessions as light-hearted and as much fun as you can when teaching a trick and your puppy will learn faster.

Finding the right training class

Attending a good training class is fun and useful. It will help you learn, correct your mistakes and

Teach the 'wave' by encouraging your puppy to paw at your hand to get the food. Once he learns to touch your hand, you can raise it high until, eventually, he is 'waving'.

A good puppy class will support your training and help with the socialization of your puppy.

keep you motivated. However, it is really important to find one that uses similar methods to those given in this book. Some instructors still use a degree of compulsion and punishment. Avoid any that encourage the use of choke-chains, American or prong-collars, or that do not advocate the use of food treats and toys for reward. Attending a bad class will set your training back and may even spoil the good relationship you have with your puppy.

I recommend that you ask the following people for details of the puppy-training or dog-training classes they recommend:

- Local vet
- Dog warden (try your local council for the phone number)
- Local rescue organizations
- The Kennel Club
- Local groomers/boarding kennels/pet shops
- Dog-owning friends.

If you ask enough people, you will end up with a list of local classes, as well as some idea of their quality. I also recommend that you contact the trainers you think may be suitable and ask to visit all of the classes *without your puppy*. It is very important to visit without your puppy so that you can make an objective assessment, without being distracted or asked to join

Take care!
- If your puppy is shy or already aggressive with people or other dogs, it would be better to train him yourself at home, or to arrange individual tuition with a good trainer, as well as getting advice from a dog behaviourist.

in. Watch the adult classes as well as those for puppies. Make an assessment based on the following criteria:

- Positive training methods for training puppies/dogs and humans, using praise, food treats and games with toys
- Training being effective for both people and puppies/dogs, so that all are learning and progressing
- A calm, ordered class
- Any off-lead play being carefully managed and supervised, with just a few puppies off-lead for short periods of time
- People and puppies/dogs having fun
- Puppies and adult dogs in separate classes; puppies under 20 weeks kept separate from adult dogs
- Dogs within a class all being at a similar level (e.g. beginner, intermediate, advanced)
- Small class sizes (eight puppies/dogs at a maximum per trainer/assistant)
- No stress and tension
- No choke-chains, prong-collars or electric collars
- No rough treatment, grabbing, shaking, shouting or pinning puppies or dogs to the floor
- No spraying with water pistols/air sprays
- No throwing of noise-makers
- No humiliation or shaming of the owners.

Finally, ask yourself, 'Would my puppy/dog and I look forward to coming to this class?' Enrol in the class you think is most suitable, even if you have to wait a while to get a place on it or have to travel a distance to get there – it will be worth it!

Requests or command?

If your puppy has been brought up using the methods given in this book, he will have a good relationship with you, will be happy to do as you say and will want to please you. If your puppy has this attitude, it is much easier and pleasanter for both of you to ask him to do things, rather than order him about.

Difficulties sometimes occur if your puppy has not been adequately proofed against distractions or if, during adolescence, he decides that he would rather do something else rather than please you. In these cases you should change your request into a command, making it louder and more forceful and increasing the urgency in your voice. A strong command, given in such a way that your puppy cannot fail to hear it, will probably produce the response you require, especially as you will not have desensitized him by using such a tone in everyday life.

If you are still ignored and you are certain that your puppy knows what you require, go back to restraining him physically with a long line so that he has no choice about complying, and practise until he becomes more responsive.

Becoming educated

The more a puppy learns, the easier he will find it to learn more. As you work with your puppy, your training skills and his ability to understand your requests will improve. He will also learn that you want him to do 'something' and will begin to try harder to find out what it is that you want. If you progress far enough with his training, you will find that you reach an almost telepathic understanding as your puppy tries to anticipate what it is that you wish him to do.

Rewards used during training will help your puppy learn to enjoy working with you.

Teaching useful tricks, such as finding your keys, will make your puppy's life more fulfilling.

CHAPTER 16
Adolescence and beyond

A very young puppy will be dependent on you for all his needs. Since you will be the centre of his world, he will follow you around and be extremely willing to please. As he grows up, the focus of his world will shift to include the rest of his environment and he will begin to develop a more independent attitude. This process will begin when he is about 18 weeks old and will build up gradually until true adolescence is reached at about six months.

At about six months of age, sexual maturity occurs and all the hormonal changes associated with this will bring further alterations to your puppy's behaviour. (Puberty can be as early as five months in smaller dogs, and seven months or later in larger breeds.)

Female dogs will come into season for the first time and quite dramatic changes in temperament can occur just before and throughout the season. Your puppy may be moody, difficult to motivate, quiet and bad-tempered during this time. She may also show competitive or aggressive behaviour towards any other female dogs in the household at this time.

Male dogs begin to cock their leg as their testosterone levels rise. They become interested in females around this time, sniffing scent marks to find out which other dogs live in the area, and begin to mark their territory in earnest. They may become more interested in competing with other males that they encounter on a regular basis, and this could escalate into scraps and squabbles at times.

Both males and females will develop more of an interest in their environment away from home. They are bigger and bolder now and will find it easy to be

Adolescence is a difficult time and most owners have a period of wondering what has 'gone wrong' with their well-behaved puppy.

At adolescence, a puppy's attention turns to the outside world and owners and their wishes become much less important.

more independent. There may be a change in the way they deal with potential threats, and they could start to use aggression as a means of feeling safe, instead of running away or using appeasement. Any behaviour problems that were lying below the surface are likely to manifest themselves at this age.

The outside world and environment will be of prime importance to your puppy at this time. What you want will seem less important to him and, while you should not give him the chance to disobey you, and you need to prevent him getting his own way all the time, be sympathetic to the fact that it is natural for his attention to be distracted.

The change in your puppy's attitude at this time can be disheartening unless you are expecting it. When very young, your puppy will be easy to teach and constantly attentive. Gradually, as he begins to mature and becomes more interested in what is going on around him, it will feel as though you are losing the closeness with him that you once enjoyed. He will be less easy to control and less willing to listen to what you want. Adolescence is a very selfish time, and what he wants will become more important to him now than what you want.

At some point in your puppy's development, you will probably feel that all the things he has learned seem to have been forgotten. Try not to worry about this too much as they will not have been forgotten, but put aside for the time being while he concentrates on

other things. Persevere with your training, patiently insisting that he does what you require.

Just a phase

It is tempting to get angry at these times, but try to remember that it is just a phase he is going through. Adolescent puppies can be exasperating and this can lead to you becoming really angry and aggressive. Feelings of failure are normal, but remember that this phase will pass and you will both emerge on the other side older and wiser. Try to end a training session before you get too annoyed. If you feel yourself getting cross, ask your puppy to do something really easy, such as 'Sit', praise him for doing this and end the session. Punishing your puppy will not help, but will only serve to put more distance between you and make the rest of his environment seem even more attractive.

A common problem at this time is that of not coming back when called while out on a walk. Since other things in the outside world now hold more interest for your puppy, he is likely to prefer to investigate rather than come back to you, no matter what rewards you are offering. If this is beginning to happen, attach a long line to his collar before letting him run free. Call him back whenever he is getting to the end of the line and use it to enforce the voice-cue if necessary (be careful not to get caught up in the line, and do not use it around children or frail people). In this way you will prevent him from learning that the

Stay in control by using a long line if necessary and keep working at the training exercises without expecting too much.

unwanted behaviour (running off) brings rewards. You will also be teaching him that he has no choice but to go back to you if he hears you call.

Just like children going through their teenage years, adolescent puppies can be difficult to live with. Not only are they becoming more interested in the world around them than they are in you, but they are also starting to assert themselves and test how far they can push you. Tighten up on your leadership role if necessary (see page 78) and try not to issue commands that cannot be enforced should your puppy decide to ignore them.

Adolescence will pass more easily if you know about it, expect it and do not worry too much about it. Luckily, it does not last as long in dogs as in humans, and by the time your puppy is 12 to 18 months old it will have passed. During the adolescent phase, it will sometimes seem that, even after all your careful early work, you now have a puppy that is more trouble than he is worth. Do not despair! Your puppy will soon mature into a dog that will, once again, look to you for decisions, be willing to please and, provided you have continued with the training, be under your control.

Young adults

Gradually your puppy will grow out of adolescence and become a young adult. At this age, all your hard work will start to pay off and you will begin to reap the rewards. He will still be 'filling out' physically and emotionally for some time, but the difficult days of adolescence are over and you can afford to relax a little now. Socialization needs to continue throughout a dog's life to an extent, but this should happen naturally as you take your well-mannered dog out and about with you. Training may need topping up from time to time, but the voice-cues you instilled early on will not be forgotten, especially if they are in constant use.

In the end, it really does work!

After all your dedication in the early months, you will eventually have a dog that is well mannered, obedient and willing to please. He can live as a member of the family and be taken anywhere, or be left at home if necessary without misbehaving. He will be well socialized with a happy, outgoing nature. He will know how to play with humans, both adults and children, and will mix easily with other dogs as well. Since you have not stifled his character by over-disciplining him, he is free to be himself without fear of being punished and his true character will shine through. He will repay the time and trouble you took during his puppyhood many times over throughout the years. He will be a friend to come home to, loved and admired by all who know him, and a dog you will be proud to own.

Once your dog matures, his focus will shift back to the family. All the hard work finally
pays off!

APPENDIX
Socialization programme

SOCIALIZATION PROGRAMME	Put a tick in the box for each encounter, entering as many ticks per box as possible.		

		7–8 WEEKS	8–9 WEEKS	9–10 WEEKS
ADULTS (MEN AND WOMEN)	Young adults			
	Middle-aged adults			
	Elderly people			
	Disabled/infirm			
	Loud, confident people			
	Shy, timid people			
	Delivery people			
	Joggers			
	People wearing uniforms			
	People wearing hats			
	People with beards			
	People wearing glasses			
	People wearing motorbike helmets			
CHILDREN	Babies			
	Toddlers			
	Juniors			
	Teenagers			
OTHER ANIMALS	Dogs – adults			
	Dogs – puppies			
	Cats			
	Small pets			
	Ducks			
	Livestock			
	Horses			
ENVIRONMENTS	Friend's house			
	Shopping centre			
	Park			
	Outside a school			
	Outside a children's play area			
	Country walks			
	Garage/car boot sale			
	Restaurant/bar/café			
	Village hall			
	Slippery floor			
	Party			
	Veterinary practice			
	Grooming parlour (if necessary)			
	Boarding kennels			
OTHER	Bicycles			
	Motorbikes			
	Skateboards			
	Pushchairs			
	Traffic			

10–11 WEEKS	11–12 WEEKS	12–13 WEEKS	13–14 WEEKS	14–15 WEEKS	15–16 WEEKS

Further reading

Gwen Bailey, *Puppy School: 7 Steps to the Perfect Puppy*, Hamlyn (2005); ISBN-10: 0600610810, ISBN-13: 978-0600610816

Gwen Bailey, *What Is My Dog Thinking?*, Hamlyn (2002); ISBN-10: 0600604233, ISBN-13: 978-0600604235

Gwen Bailey, *Dogs Behaving Badly*, HarperCollins (2007); ISBN-10: 0007244371, ISBN-13: 978-0007244379

Suzanne Clothier, *Bones Would Rain from the Sky: Deepening our Relationships with Dogs*, Warner Books, reprint (2005); ISBN-10: 044669634X, ISBN-13: 978-0446696340

R. Coppinger and L. Coppinger, *Dogs: A startling new understanding of canine origin, behaviour and evolution*, Prentice Hall & IBD (2001); ISBN-10: 0684855305, ISBN-13: 978-0684855301

John Fisher, *Think Dog!: An owner's guide to canine psychology*, Cassell Illustrated, new edition (2003); ISBN-10: 1844031209, ISBN-13: 978-1844031207

John Fisher, *Why Does My Dog ...?*, Souvenir Press, new edition (1999); ISBN-10: 028563481X, ISBN-13: 978-0285634817

John Fisher, *Dogwise: The natural way to train your dog*, Souvenir Press, reissue (1992); ISBN-10: 0285631144, ISBN-13: 978-0285631144

Patricia McConnell, *The Other End of the Leash: Why we do what we do around dogs*, Random House USA, reprint (April 2003); ISBN-10: 034544678X, ISBN-13: 978-0345446787

Karen Pryor, *Don't Shoot the Dog: The new art of teaching and training*, Ringpress Books, 3rd revised edition (2002); ISBN-10: 1860542387, ISBN-13: 978-1860542381

Pam Reid, *Excel-erated Learning*, James & Kenneth Publishers, US (1996); ISBN-10: 1888047070, ISBN-13: 978-1888047073

Turid Rugaas, *On Talking Terms with Dogs: Calming signals*, Dogwise Publishing, 2nd edition (2005); ISBN-10: 1929242360, ISBN-13: 978-1929242368

D. Weston, *Dog Training: The gentle, modern method*, Gazelle Book Services (2003); ISBN-10: 1855860023, ISBN-13: 978-1855860025

D. Weston & R. Ross, *Dog Problems: The gentle, modern cure*, Gazelle Book Services (1992); ISBN-10: 185586004X, ISBN-13: 978-1855860049

Useful addresses

If your puppy has a behaviour problem, contact:

Association of Pet Behaviour Counsellors
PO Box 46
Worcester
WR8 9YS
England
Telephone: 01386 751151
Email: info@apbc.org.uk
www.apbc.org.uk

For a good-quality puppy training class in your area and advice about new puppies, contact Puppy School (founded by Gwen Bailey):

Puppy School
PO Box 186
Chipping Norton
OX7 3XG
England
Email: info@puppyschool.co.uk
www.puppyschool.co.uk
www.dogbehaviour.com

For a training class for older dogs, contact:

Association of Pet Dog Trainers (UK)
PO Box 17
Kempsford
GL7 4WZ
England
Telephone: 01285 810811
Email: APDToffice@aol.com
www.apdt.co.uk

For details of local training classes or information about new puppies, contact:

The Kennel Club (UK)
1–5 Clarges Street
Piccadilly
London
W1J 8AB
England
Telephone: 0870 606 6750
www.thekennelclub.org.uk

Index

Page numbers in *italics* refer
to illustrations.

A

adolescence 38–9, 196–8
aggression:
    in car 122–3
    chase 126–7
    dominance: towards humans 126
    during adolescence 197
    fear-induced:
        towards humans 119–20
        towards other dogs 127
    food 123–4
        exercise *122–3*
    inter-male 129
    pain-induced: towards humans
        126
    possession 123–5.
    preventing 114–29
    punishment encouraging 90
    reasons for 118–19
    territorial 120–2
    to strangers 18
    towards children 120
    towards humans 119–20
        dominance 126
    towards other dogs 127–9
    see also biting
agility course 112, *112*
alone: learning to be 35, 154–5
ambitions 86
ankles: play-biting 117–18

B

bad experiences:
    overcoming 66–8
    protection from 127
baring teeth: as warning 118
barking:
    at arrival of visitors 120, 122
    excessive 146–8
    in playpen 46–7
bed: positioning 155
begging: avoiding 149–50
behaviour 78–91

bad: preventing 78–9
good:
    encouraging 78–9
    rewards for 78
problems 91
    showing up in adolescence
    197
behaviourist: advice from 193
biting:
    preventing 114–19
    reasons for 118–19
    warning of 118
Bloodhounds *11*
body language 24–5, *25*, *26–8*, 168
body-maintenance needs 80–1
bones:
    aggression over 124–5
    exercise *124–5*
Border Collies 12–13
boundary-setting 81–5
Boxers 12–13
breeders:
    buying puppy from 16–17
    role in socialization 16, 56
breeds 10–15
    characteristics 12–13
    choosing 14
    groups 10
    show stock 14–15
    working strain 14–15
bull breeds: games for 96
buying puppy:
    choosing 17–18
    sourcing 16–17

C

car:
    aggression in 122–3
    leaving puppy in 133, 153
    travel in 151–3
castration: to prevent aggression 129
cats:
    introducing to puppy 42
    socializing with 64
Cavalier King Charles Spaniels 12–13
challenges: winning 86

chasing 103–5
    aggression induced by 126–7
    elementary chase recall: exercise
        104–5, 105
    games *126*
    waiting for permission 103–5, *103*
chewing 38, 80, 130–3
    reasons for 130
    stages of 130–1
    teaching right and wrong 132
chews:
    aggression over 124–5
        exercise *124–5*
    hiding food in 131
    objects suitable as 131–2, *131*
children:
    aggression towards 120
    bringing up puppy with 31–3
    greeting puppy 144
    handling puppy 134
    introducing puppy to 40–1
    playing with puppy 99–100, *100*
        preventing jumping up 145
    retreat from 47–8
    socializing with 59, 62
    teaching puppy 32
    teasing by 120
choosing puppy:
    breed 14
    individual 17–18
classes: finding 192–4
clothing: play-biting 117–18
Cocker Spaniels *11*, 12–13
collar:
    care with 176
    introduction to 51–3
Collies *10*, *11*
    Border 12–13
colour spectrum: visible *24*
colour vision 23–4
commands *see* voice-cues
communication:
    canine 19
        classes 62–4
    by leader 85
concentration: building 170–1

confidence 17, 118–19
contests 85–6
correction 88–9, 90
   by older dog 35
crawl: teaching 190

D
Dachshunds 11
Dalmatians 11
dancing 112
Deerhounds 11
delivery people 120–2
development: stages 36–9
discipline:
   correction 88–9
      by older dog 35
   physical punishment 90–1
   for toilet accidents 74
Dobermanns 11
dogs:
   adult 198
      in family 33–5
      introducing puppy to 41–2
      socializing with 62–4
      teaching puppies 64
   aggression towards 127–9
   rough play with 127–9
dominance aggression: towards
      humans 126
down command 180, 180–1
   at distance 187–8, 188–9

E
ears: examining 137
English Springer Spaniels 12–13
exercise 80, 81
exercises:
   aggression 122–5
   chase recall 104–5, 105
   down 180, 180–1
      at a distance 187–8, 188–9
   drop command 106–7, 107
   elementary chase recall 104–5, 105
   grabbing 139–41, 139
   greeting people 144–5
   grooming 138
   handling 136–7, 137
   off command 88–9
   recall 104–5, 105, 172–3
   retrieve 98–9, 99
   roll over 184, 184–5

sit 178–9, 178–9
stand 182, 182–3
stay 186–7, 186–7
taking food gently 148–9
tug-of-war games 106–7, 106–7
walking on lead 174–5, 174–5
exploring:
   phase for 130–1
   providing objects for 131–2
eyes 23
   examining 137
   opening after birth 36

F
facial awareness 29
family:
   bringing puppy into 30–5
   with second dog 33–5
fear 68
   aggression induced by 119–20
   of new experiences 55
   of new things 37, 38
   of strangers 18
feeding: small meals 80
feet: play-biting 117–18
females: in season 196
'Find it' game 109
flyball 112, 113
food:
   aggression over 123–4
   taking gently 148–9
      exercise 148–9
   as treats in training 158, 159–60
frustration: dealing with 142–3

G
games see play
genes 10–11
German Shepherd Dog 12–13, 18
Golden Retrievers 12–13, 17
grabbing: exercise 139
greeting people 144, 145–6
   exercise 144–5
grooming 138–41
   exercise 138
growling: as warning 118
guarding dogs: games for 96
gundogs 10, 189
   chewing 130
   games for 96
   working tests 113

H
habits:
   bad: preventing 142
   good 142–53
habituation 54
   to situations and objects 65–6
hand signals 168–9, 168
handling 134–7
   by children 134
   exercise 136–7, 137
   getting puppy accustomed to
      135–7
   grooming 138–41
   by other people 141
   preparation for vet's examinations
      135–7
   by strangers 60–1, 141
head: sensitivity of 138, 140
hearing 22
   development 36
herding dogs 189
   games for 96, 103
home:
   arrival of puppy 40–5
   first night 43–5
horses: socializing with 64, 65
hounds 10
   games for 96, 103
house-line: using 116, 116
housetraining 70–7
   accidents: cleaning up 72
   differentiating between nest and
      toilet area 37
   in flats 76–7
   lapses 77
   at night 44, 45, 73–4
   punishment 74
   signs to look for 70–1, 72
   time required to learn 74
   times for 70
   on voice-cue 74–6
   words to use 70, 71

I
independence 38
inherited traits 11, 15, 18
isolation:
   during day 44
   first night at home 43–4
   from mother 37
   learning to tolerate 35, 154–5

J

Jack Russell Terrier 12–13
jumping: avoiding 113, 189
jumping up 143–5

K

kennels: indoor 48–9, *48*

Labradors 12–13, *14*, 101, 130
lead:
    introduction to 53
    trailing 79
    walking on 173–7
        exercise 174–5, *174–6*
leadership 81–5, 90
    challenging for 126
    enforcing during adolescence
      198
    by owner 78, 120
    training dependant on 157
'Leave' command 105, *105*
lifespan 54–5
lifting *136*, 137
litter:
    isolation from 37, 43–4
    siblings from 34–5
livestock: becoming accustomed to
    64
looming over puppy 139–40

M

'mad five minutes' 116–17, *117*
males:
    aggression between 129
    marking territory 196
manners: good 142–53
maturity 39, 196
meal times: humans' 149–50
mental development 37
mongrels 11
mother:
    age for puppy to leave 18–19
    assessing puppy with 17, 18
    education from 19
    separation from 37, 43–4
mouth:
    examining *137*
    using 29
mouthing 38
movement: detecting 23–4
muzzles 141

N

nails: cutting 136
name: responding to 171–3
names of objects: teaching 110
needs: meeting 79–81
nervous aggression 18, 19
nervousness: with strangers 17–18

O

obedience trials 112
'Off' command *88–9*
owners:
    living alone 31
    puppies reflecting characters of
      30–1

P

pack leader 82, 83–5
    see also leadership
pain: aggression induced by 126
pastoral dogs 10
paws: handling *136*
pet shops 16
pets:
    puppy: with other species 35
    second dog 33–5
    from same litter 34–5
play 80, 81, 92–113
    chase games 103–5
    with children 99–100, *100*
    competition in 101
    controlling 102–3
    developing bond through 93
    at exciting times of day 101–2
    fetch the slipper 110
    'Find it' game 109
    fun in 100–1
    grown-up games 112–13
    human games 92
    importance of 92–4
    learning through 96
    with litter-mates 37
    playing to win 86
    preventing jumping up during 145
    retrieving game 110
    as reward for good behaviour 94
    rough: with other dogs 127–9
    shake-and-kill games 108–9,
      108–9
    as training rewards 160–1
    tug-of-war games 106–7, *106–7*

play-biting 37, 38, 47
    alternatives to 96–8
    excessive 114–18
    redirecting 114
    walking away from 114–16
play-fighting 85–6, 117
playpen 42, 44, 45–8, *45*, *47*
police dogs 112
positions:
    down 180, *180–1*
      at a distance 187–8, *188–9*
    roll over 184, *184–5*
    sit 178–9, *178–9*
    stand 182, *182–3*
    stay 186–7, *186–7*
possession aggression 123–5
prey-killing: play at 37
puberty: age reached 196
punishment:
    after absence 155
    physical 90–1
    resulting in aggression 119–20
    for toilet accidents 74
    in training 157–8
puppy farms 16–17

R

recall training 171–3
    elementary *104–5*, 105
    exercise *172–3*
    reinforcing in adolescence 197–8
reprimands 88–9
resources: controlling 84–5
respect 81, 82
resting place 80
restraint 53
    gentle 135
retrieve:
    exercise *98–9*, 99
    in water 113
Retrievers 12–13, *17*
rewards:
    games as 94
    for good behaviour 78
    jackpots 164
    motivation and 158–61
    random 163–4
    timing 164–7
    in training 156–62
    types 158–61
routine: establishing 60–1

S

safety 79
scent: importance of 22–4
season: changes during 196
self-confidence 118
separation:
    coping with 35, 154–5
    during day 44
    first night at home 43–4
    from mother 37
settling down 150–1
sexual maturity 38, 196
shake-and-kill games 108–9, *108–9*
Sheepdogs 12–13, 15
shyness 68–9, *119*
sight 23–4
    development 36
sit 178–9
    exercise *178–9*
sleep 80–1
smell, sense of 20–2
    development 36
social needs 79
socialization 16–17, 29, 54–69
    developmental stages 36–9
    getting used to situations and
        objects 65–6
    with humans 60–2
    with other animals 35, 64–5
    with other dogs 33, 127
    poor: resulting in aggression 119
    programme 200–1
    in single-person household 31
    successful training 56–9
    training classes 62–4, 193, *193*
Spaniels:
    Cavalier King Charles 12–13
    Cocker 11, 12–13
    English Springer 12–13
    Springer 12–13, *15*
spoken word: responding to 24–5
Springer Spaniels 12–13, *15*
Staffordshire Bull Terrier 12–13,
    94
stair gate 42, 49–50, *49*
stand command 182, *182–3*
stress:
    effects on toilet training 77
    mild: benefits of 36
submission: urination as sign of
    77

T

tail-wagging game 170–1
teasing 120
teeth:
    baring: as warning 118
    examining 136
teething 36, 38, 130
temperament 11
Terriers 10
    games for *95*, 96, 101
    Jack Russell 12–13
    Staffordshire Bull 12–13, *94*
    West Highland White 12–13
    Yorkshire 12–13
territorial behaviour 38
    aggression 120–2
toilet training *see* housetraining
toy dogs 10
toys 96
    aggression over 125
    games with 94
    keeping control of some 86, 98–9
    suited to breeds 94–6
    suited to games 96
    as training rewards 160–1
training 156–95
    classes: finding 192–4, *193*
    commands and signals 168–9,
        168
    distractions 163
    down command 180, *180–1*
        at a distance 187–8, *188–9*
    enjoyment 167–8
    housetraining 70–7
    learning with other person 165–7
    leave command 105, *105*
    length of sessions 167
    place associations 161–3
    positions 178–88
    programme for 169
    recall *104–5*, 105, 171–3, 197–8
    relationship for 157–8
    requests and commands 194
    reward-based 156–62
    roll over 184, *184–5*
    sit command 178–9
    socialization classes 62–4
    stand 182, *182–3*
    stay 186–7, *186–7*
    success 169–70
    time required 169
    timing 164–7
    tricks 188–92
    walking on lead 173–7
    see also voice-cues
travel cage 48–9, 152, *153*
tricks: teaching 188–92, *190, 191, 192,
    195*
tug-of-war games 106–7, *106–7*

U

upbringing 30–1
urination: submissive 77
utility dogs 10

V

vaccinations 56
vet: accustoming puppy to handling
    by 135–6
voice-cues 156, 168–9
    for control 102
    'Drop' command *106–7*, 107,
        108–9, *108–9*
    'Fetch' command 104–5, 110
    'Find it' 109
    for games 102
    learning 91
    'Leave' command 105, *105*
    'Off' command 86–8
        exercise *88–9*
    responding to 161–3
    retrieve: exercise *98–9*
    for toilet training 70, 71, 74–6

W

walking: restricting distances 94
walking past people and dogs
    145–6
water: work in 113
weaning 37
West Highland White Terrier 12–13
whiskers 29
wild: life in 55
wolves 20
working dogs 10
    channelling instincts 93
    as pets 15
working trials 112–13, *113*
worming 62

Y

Yorkshire Terrier 12–13

Acknowledgements

Author acknowledgements

My heartfelt thanks go to those people who helped me and taught me so much about dog behaviour early in my career, particularly John Rogerson, the late John Fisher, Ian Dunbar, Tony Orchard and many others who contributed so much. Their input enabled me to be ahead of my time with knowledge of positive methods, and for that, and for their inspiration, I will always be enormously grateful.

Photographic acknowledgements

Special photography
Octopus Publishing Group Limited/Russell Sadur.

Other photography
Alamy/Adrian Buck 34; /Arco Images 10.
Ardea/John Daniels 11, 16.
Octopus Publishing Group Limited 129; /Angus Murray 118; /Ray Moller 12 picture 1, 12 picture 2, 12 picture 11, 12 picture 4, 12 picture 5, 12 picture 6, 12 picture 9, 12 picture 3, 12 picture 12, 39 right; /Rosie Hyde 36, 113 top; /Russell Sadur, 64, 76, 140; /Steve Gorton 12 picture 10, 12 picture 7, 12 picture 8, 43, 193.
The Kennel Club Picture Library 112, 113 bottom.

Publisher acknowledgements

The publishers would like to thank the following people and their dogs for taking part in the photoshoot: Emelye Allen and Rossi; Elizabeth, Phoebe and Tom Barley and Ralphie; Sue and Sophie Belcher and Ella; Bobs Broadbent and Nellie and Lola; Carin and Tim Browne and Henry; Geoff Burnand and Monty; Kate and Eleanor Catlin and Toffee; Natalie, William and Carrie Charles and Monty; Heather Cooper and Luey and Roxy; Denise and Liam Kavanagh and Archie and Jock; Linda Lee and Byron; Alison Littlewood and Willow; Sylvie, Angus and Krystyna Mawer and Daisy; Roz Mortimer and Amber; Kate Perry and Tess; Katy, Georgina and Michael Renny and Bonnie; Philip Robinson and Sacha; Amba Suppel and Winnie; Emma Vagg and Damian Hart and Harry. Special thanks to Bobs Broadbent, Emma-Clare Dunnett and Sue Ottmann for all their help in finding puppies and people.

Executive editor Trevor Davies
Editor Fiona Robertson
Design manager Tokiko Morishima
Designer Mark Stevens
Illustrator Sudden Impact Media
Production manager Ian Paton